T0181987

China Internet Development Report 2019

Chinese Academy of Cyberspace Studies

China Internet Development Report 2019

Blue Book for World Internet Conference,
Translated by CCTB Translation Service

 中国工信出版集团  Springer

Chinese Academy of Cyberspace Studies
Beijing, China

ISBN 978-981-33-6932-0   ISBN 978-981-33-6930-6 (eBook)
https://doi.org/10.1007/978-981-33-6930-6

Jointly published with Publishing House of Electronics Industry
The print edition is not for sale in China Mainland. Customers from China Mainland please order the
print book from: Publishing House of Electronics Industry.

This Springer imprint is published by the registered company Springer Nature Singapore Pte Ltd.
The registered company address is: 152 Beach Road, #21-01/04 Gateway East, Singapore 189721,
Singapore

Preface

The year 2019 marks the 50th anniversary of the birth of the Internet and the 25th anniversary of China's full-function access to the global Internet. Over the past year, China has made a series of progress and achievements in Internet development by following the trend of the times and further advancing the cause of building China's strength in cyberspace. *China Internet Development Report 2019* (hereinafter referred to as the "Report") is intended to faithfully record the history of China's Internet development, reflect its general situation, summarize practical experience, and present China's accomplishments in Internet development, which enables readers to understand and study China's Internet development by providing abundant materials and detailed data. The Report mainly has the following three characteristics:

(1) By linking the past, present and future, it reveals profoundly the historic opportunity brought by Internet development to the great rejuvenation of the Chinese nation. China's Internet has gone through 25 years of extraordinary history. The Report starts with a special chapter titled "Twenty-Five Years of China Internet Development", which reviews the history of China's Internet development for the past 25 years, provides a panorama of new theories and practices of Internet construction, application and management, and in particular fully displays the historic achievements and reforms of China's Internet development under the guidance of the important thought of General Secretary Xi Jinping on building China's strength in cyberspace since the 18th National Congress of the CPC. Standing at a new historical starting point, the Report adheres to the guidance of Xi Jinping Thought on Socialism with Chinese Characteristics for a New Era, particularly his thought on building China's strength in cyberspace. With a view to realizing the great rejuvenation of the Chinese nation, it forecasts the future of China's Internet development and hopes to fully reflect Chinese people's confidence and determination in grasping the historic opportunity of information revolution and accelerating the progress toward the strategic goal of building China's strength in cyberspace.

(2) The Report comprehensively assesses the progress of Internet development in all provinces, autonomous regions and municipalities directly under the Central Government of China (excluding Hong Kong, Macao and Taiwan), playing a

demonstrative, guiding and incentive role. China has released the Internet Development Index since 2017. In 2019, China adjusted and optimized some Internet development indexes according to Internet development and actual situation on a consistent basis with the previous versions on the whole. The comprehensive evaluation system was also further improved to verify data sources on a scientific basis. The Report evaluates the comprehensive Internet development of 31 provinces, autonomous regions and municipalities directly under the Central Government (excluding Hong Kong, Macao and Taiwan), aiming to provide a comprehensive, objective and accurate overview of their level of Internet development. In particular, it also releases top 10 single-item rankings in 6 fields information infrastructure construction, innovation capacity, digital economy, Internet application, cybersecurity and cyberspace governance, which, as a model, guides the accurate judgment of comparative advantages in different areas and encourages Internet development across China.

(3) The Report provides a panorama of new accomplishments, experiences and trends in terms of China's Internet development based on practice-oriented, problem-oriented and goal-oriented principles. China's innovative practice of Internet development in 2019 provides abundant material, vivid case and solid foundation for the compilation of the Report. Putting emphasis on the practices of China's Internet development, the Report showcases an overall picture of new developments, achievements and trends of Internet development in 2019 in eight key fields, namely, information infrastructure construction, network information technology, digital economy, digital government and e-governance, web content construction and management, cybersecurity, legal system of cyberspace, and international cyberspace governance. It also systematically presents China's up-to-date experience and innovative practices in building China's strength in cyberspace.

We hope the compilation and publication of the Report will give fresh impetus to China's Internet development, provide a new window for readers to learn about China's Internet development from a new angle, and guide Internet development in other countries with its experience. This is also our original intention and mission of compiling the Report.

Beijing, China   Chinese Academy of Cyberspace Studies
September 2019

Twenty-Five Years of China's Internet Development

As one of the greatest inventions of the twentieth century, the Internet is profoundly transforming people's production and life and forcefully promoting economic and social development. It is unparalleled to any other scientific and technological achievements in terms of the speed of development, scope of influence and depth of impact. In 1994, China achieved its full-function access to the Internet and became the 77th country in the world to get access to the Internet. Since then, China's development has become closely intertwined with the Internet and China has got closer to the world. After 25 years of magnificent practices, China has rapidly narrowed its information gap with developed countries and has become a country with an extensive cyber presence that captures the attention of the world. Especially since the 18th National Congress of the CPC, China has formulated an important strategic plan of building its strength in cyberspace and gradually embarked on a path of Internet development with Chinese characteristics, thus opening a new chapter of China's Internet development and governance. With its wide penetration into economic, political, cultural, social and other fields, the Internet has profoundly changed the country's appearance, people's living conditions and China's status in the world, provided valuable historic opportunity to the great rejuvenation of the Chinese nation.

Arising in Accordance with the Situation: Internet Development Thriving in China

Since its birth, the Internet has impacted unprecedentedly various aspects, including an extensive change in the forces and relations of production, significant impact on social and economic development and the production and life of human beings, as well as profound adjustment in international, political and economic structures. In addition, it has also posed unprecedented risks and challenges to national security, and unprecedented shock on the exchange, integration and confrontation among different cultures and values. As a latecomer, China has only linked up with international Internet for 25 years. But over these years, it has taken the great opportunity of

Internet development to advance the cause of socialist modernization with Chinese characteristics. It has conformed to the tides of information revolution, grasped the rules of Internet development, and correctly handled the relationships between security and development, openness and self-reliance, management and service. It has made sound and rapid progress in Internet development that attracts everyone's gaze.

(1) The 25 years of China's Internet development have been 25 years of determined reforms and pioneering efforts. China's Internet development started under the historical background of reform and opening-up and further drove them. Over the past 25 years, China has proactively created favorable policy, legal and market conditions for Internet development, constantly improved the leadership for the management of the Internet, and made great efforts to promote the Internet's rapid development and wide application in China by upholding the spirit of reform and innovation. Through proactive and pioneering measures, new breakthroughs and new miracles have been constantly achieved in Internet development.

(2) The 25 years of China's Internet development have been 25 years of leading-edge development and transformation. To promote Internet development, China has acted in strict accordance with the National Strategy for Innovation-driven Development, given full play to the driving and leading role of informatization, and taken innovation in information technology and industrial progress as the motive force. Over the past 25 years, China has made a series of achievements on technological innovation in the information field. Breakthroughs have been made in the research of 5G, high-performance computing, quantum communication and other fields. The world's largest Internet infrastructure has been completed. It has many indicators ranking first in the world, including the coverage of fixed fiber optic networks, the scale of 4G network, the number of broadband users and the number of Internet users. The Internet is accessible by thousands of millions of households. A great number of Chinese enterprises have taken the lead in the world. With its scale ranking second in the world, digital economy has become a new driving force and highlight of China's economic development and has driven the high-quality economic development in China in an all-round way.

(3) The 25 years of China's Internet development have been 25 years of serving the community and benefiting the people. China has always followed the fundamental tenet of benefiting nearly 1.4 billion Chinese people in Internet development and has been committed to keeping up with people's ever-growing needs for a better life by developing the Internet. Over the past 25 years, the Internet has been integrated into every aspect of social production and social life, becoming a new space for people to study, work and live. It has greatly enhanced the country's national governance and social governance capabilities, and improved its public service level in education, culture, transportation and health care. As the main channel for people to produce, disseminate and obtain information, the Internet has fostered wider-range transmission and

sharing of ideas, cultures and information, and greatly enriched people's spiritual and cultural lives. Cyberspace has become the spiritual home of hundreds of millions of people where they feel more fulfilled, happier and safer in sharing the achievements of Internet development.

(4) The 25 years of China's Internet development have been 25 years of law-based governance and guaranteed security. Cybersecurity has always been a major issue born with the Internet. Over the past 25 years, China has strictly followed law-based governance and the bottomline of cybersecurity for Internet development. Through legal regulation, administrative supervision, industrial self-discipline, technological guarantee, public supervision and social education, all social forces have been fully mobilized to jointly maintain cybersecurity and order in cyberspace, keep purifying network ecosystem, effectively respond to cybersecurity threats, comprehensively promote the legalization of cyberspace, resolutely crack down on illegal and criminal activities on and via the Internet by law, safeguard people's legitimate rights and interests, and ensure long-term healthy and orderly development of the Internet.

(5) The 25 years of China's Internet development have been 25 years of openness, cooperation, mutual benefit and win-win relationship. The Internet has turned the world into a global village and made the international community a highly interdependent community with a shared future. Over the past 25 years, China has kept in mind both its domestic and international imperatives, established in-depth international exchange and cooperation on cyberspace, and absorbed technology, personnel, capital and management resources from all over the world to push forward its Internet development. As an active participant in international cyberspace governance, it has played an important part in fostering a more equitable and reasonable global Internet governance system, and contributed China's wisdom and experience to global Internet development and governance.

Planning in Accordance with the Situation: Accelerating Progress Toward the Strategic Goal of Building China's Strength in Cyberspace

Since the 18th National Congress of the CPC, General Secretary Xi Jinping has attached great importance to cyberspace affairs and proposed the strategic goal of building China's strength in cyberspace. Staring from the development trend of information age and the overall situation at home and abroad, Xi Jinping has put forward a series of new concepts, new ideas and new strategies for cyberspace affairs based on the practices of Internet development and governance in China, and elaborated on major theoretical and practical issues concerning cybersecurity and informatization systemically, which together constitute his important thought on building China's strength in cyberspace. General Secretary Xi Jinping's important thought on building China's strength in cyberspace has clarified the important role that cyberspace affairs play in the overall cause of the Party and the state, and clearly defined the strategic objectives, principles and requirements, international views and fundamental methods. It is the "cyberspace part" of Xi Jinping Thought on Socialism with Chinese Characteristics for a New Era. It constitutes a scientific summary and theoretical refinement of the path to socialist governing of the Internet with Chinese characteristics. It is an ideological guide and follow-on action path to lead the development of cyberspace affairs. It plays a greater role in China's cyberspace affairs, and guides China to actively adapt to and take the lead in the new round of scientific and technological revolution and industrial transformation. As a result, China accelerates its march from a big network country to a cyberpower and makes historic achievements and changes in cyberspace affairs.

Under the guidance of the important thought, China has strengthened the overall planning and coordination and top-level design, pressed ahead the reform of the leadership system for Internet management, and has set up the Central Cyberspace Affairs Commission. In 2018, the Group was renamed as the Central Cyberspace Affairs Commission in order to strengthen the comprehensive coordination over major issues such as construction and management of web contents, cybersecurity, informatization development and global cyberspace governance. It has released a series of strategic and institutional documents, and improved the systems and institutions for cyberspace administration, thus creating a strong synergy around the whole nation to advance the cyberspace affairs.

Under the guidance of the important thought, China has strengthened the construction and management of online contents by adhering to the principle of "maintaining positive energy, keeping things under control and correctly utilizing the Internet". It has actively fostered and practiced socialist core values and vigorously developed a positive cyber culture, aiming to create a sound network ecosystem for the public, build a concentric circle online and offline, and provide spiritual impetus and cultural support for realizing the great rejuvenation of the Chinese nation.

Under the guidance of the important thought, China has focused on building a cybersecurity guarantee system, and constantly strengthened the protection of critical information infrastructure security and data security. By formulating and implementing *Cybersecurity Law of the People's Republic of China* (hereinafter referred to as "Cybersecurity Law"), it aims to raise people's awareness and skills of cybersecurity protection and has built a solid foundation for cybersecurity. These acts have effectively safeguarded the security and interests of national cyberspace.

Under the guidance of the important thought, China has given full play to the leading role of informatization to speed up core technological innovation in the information field and vigorously develop digital economy. It has facilitated the deep integration of the Internet and real economy, made information services more convenient and beneficial for people, and further advanced the "Internet+" action. Efforts have also been made in promoting the healthy development of e-government and comprehensively pushing forward Internet-assisted poverty alleviation so that the Internet better benefits the society and people.

Under the guidance of the important thought, China has strengthened international exchange and cooperation on cyberspace, and has acted on the "four principles" for advancing the reform of the global Internet governance system and the "five proposals" for building a community of shared future in cyberspace proposed by General Secretary Xi Jinping. By actively participating in and leading international cyberspace governance and fostering a community with a shared future in cyberspace, China has continuously strengthened its power of discourse and influence on cyberspace on a global stage.

Acting in Accordance with the Situation: Grasping Opportunities for Informatization

Looking at the history of social development, we can see that having experienced agricultural and industrial revolutions in the past, the human society is now in the midst of information revolution. Agricultural revolution enhanced the capacity of humans to survive, marking the progression of humanity from hunting and gathering to farming and herding, from a primitive society to a civilized one. Industrial revolution expanded the limits of human labor, replacing physical labor with machinery and manual workshops with large-scale factories. Now information revolution boosts human intellect, causes another leap forward into productive forces, accelerates the flow and sharing of resources such as labor, capital, energy and information, and exerts a profound impact on international politics, economy, culture, society, ecology and military affairs.

China was once a leading economic power in the world in the agricultural age. However, when Europe experienced the industrial revolutions and the world underwent a period of profound changes, China missed out on the historic opportunity to develop in step with the rest of the world and became vulnerable to attacks. After the Opium War, in particular, the Chinese people sunk further into the miserable plight of poverty and weakness, being trampled over at will. Building a modern socialist powerful country that is strong, democratic, culturally advanced, harmonious and beautiful and the realization of the Chinese nation's rejuvenation have been the greatest dream of the Chinese people since modern times. This embodies the highest and the most fundamental interest of the Chinese nation. Through the efforts of several generations, in just a few decades, China has completed a course that took developed countries several hundred years and finally caught up with the times in big strides. It has created a miracle in the history of the human society. With socialism with Chinese characteristics entering a new era, the Chinese nation has achieved a tremendous transformation: it has stood up, grown rich and become strong. China is now closer to the Chinese Dream of the great rejuvenation of the Chinese nation, and is more confident and capable of achieving it than at any other time in history.

Now, as the timeframes of the great revolution of science and technology in information age and the great rejuvenation of the Chinese nation are converging, new contents have been added to China's development in an important period of strategic opportunity. This is an important historical opportunity for the Chinese nation. As a new round of technological and industrial revolution is gaining momentum, under the scientific guidance of Xi Jinping Thought on Socialism with Chinese Characteristics for a New Era, particularly his thought on building China's strength in cyberspace, and on the strong basis of the historic achievements of China's Internet development in the past 25 years, China will surely seize the historic opportunity of the information revolution for Internet development with the concerted efforts of more than 800 million Internet users and nearly 1.4 billion people in China. It will accelerate the progress toward the strategic goal of building China's strength in cyberspace and make due contribution to the realization of the Two Centenary goals and the great rejuvenation of the Chinese nation.

Overview

At present, modern information technologies represented by the Internet, big data and AI change with each passing day; a new round of technological and industrial revolution gains momentum, and the pace of digitization, networking and intelligentization speeds up. They together have facilitated a new leap into social productivity and improved the ability of human beings to understand and change the world in a wider scope, at a deeper level and with higher standards. As the timeframes of the great revolution of science and technology in the information age and the great rejuvenation of the Chinese nation are converging, new contents have been added to China's development in an important period of strategic opportunity. China enters a new stage of rapid development of cyberspace affairs.

The year 2019 is the 70th anniversary of the founding of the People's Republic of China and a crucial year of building a moderately prosperous society in all respects. It also marks the 50th anniversary of the birth of the Internet and the 25th anniversary of China's full-function access to global Internet. Over the past 25 years, China has kept pace with times based on its basic national conditions and actively absorbed and has drawn on the experience of other countries in the world of Internet development. On this basis, China's Internet has grown increasingly stronger and scored great achievements that have captured the world's attention. Especially since the 18th National Congress of the CPC, the Party Central Committee with Comrade Xi Jinping at its core has attached great importance to the cyberspace affairs. It has set clearer goals of building China's strength in cyberspace, taken more forceful measures and made more solid steps in Internet development. As the cyberspace affairs become more important in the overall cause of the Party and the country, historic achievements and reforms have been made in their development. At present, China's cyberspace affairs enter a new stage of overall and accelerated development after crossing the initial stage and basically finishing the foundation work. In 2019, faced with a new situation and a new task in the new stage of Internet development, China's Internet development has adhered to the guidance of Xi Jinping Thought on Socialism with Chinese Characteristics for a New Era, particularly his thought on building China's strength in cyberspace, by constantly strengthening construction and management of online contents, enhancing its capacity to protect cybersecurity, and giving play to the driving and leading role of informatization. It has also actively participated in international governance and cooperation on cyberspace. The construction of building China's strength in cyberspace, digital China and smart cities has progressed at an accelerated rate, which has made positive contributions to and provided strong support for the progress of the cause of the Party and the country.

I. New Achievements and Progress of China's Internet Development in 2019

In 2019, under the guidance of Xi Jinping Thought on Socialism with Chinese Characteristics for a New Era, particularly his thought on building China's strength in cyberspace, China has firmly grasped the historic opportunity of informatization development to accelerate network infrastructure development and improve its capability of independent innovation in network information technology. It has also vigorously developed digital economy, made information services more convenient and beneficial for people, strengthened construction and management of web contents and focused on enhancing its capacity to protect cybersecurity. Other efforts were made in deepening international exchange and cooperation on cyberspace. With these acts, a series of new achievements have been made in Internet development.

1. Optimized and Upgraded Network Infrastructure

General Secretary Xi Jinping stressed that China should strengthen information infrastructure construction and focus on investing in next-generation information infrastructure. It should accelerate the building of high-speed, mobile, secure, and ubiquitous next-generation information infrastructure and improve the intelligence level of traditional infrastructure. An infrastructure system that caters to the needs of intelligent economy and intelligent society should be established. By keeping abreast of the trend of informatization development, China has continued to promote information infrastructure construction, accelerated the implementation of the "Broadband China" strategy, and furthered the implementation of the "Boosting Internet Speed and Lowering Internet Charges" policy. It has also constantly strengthened the construction of high-speed broadband networks and continuously optimized its network structure. Its network performance has been improved significantly. Its quantity of critical resources for the Internet ranks top in the world. More than 90% of broadband users in China use optical fiber access networks, ranking first in the world. As of June 2019, China's optical fiber access users reached 396 million, accounting for 91% of the total number of broadband Internet users.[1] China's fixed

---

[1]Data Source: The Ministry of Industry and Information Technology of the People's Republic of China (MIIT).

broadband Internet users continue to migrate to high speeds. The proportion of broadband users with access rates of 100 Mbps and above has steadily increased, and the total number of fixed broadband Internet users has reached 435 million. Mobile Internet has developed rapidly, with a total of 7.32 million mobile communication base stations, with 4G base stations accounting for 60.8% or 4.45 million units.[2] On June 6, 2019, China's 5G licenses for commercial uses were officially granted, marking China's official entry into the first year of 5G's commercial use. Its quantity of critical resources for the Internet has increased sharply. IPv6 scale deployment has accelerated. China's domain name system has been improved. As of June 2019, the number of IPv4 addresses has reached 385.98 million,[3] and that of IPv6 addresses has reached 50,286/32s, an increase of 14.3% since the end of 2018, ranking first in the world; the number of IPv6 active users has reached 130 million, and the number of users with IPv6 addresses allocated by China's basic telecommunication companies has reached 1,207 million.[4] Abundant IP address resources have provided great support for China's rapid Internet development. Besides, as of June 2019, China had 48 million domain names. Among them, the number of ".CN" domain names was 21.85 million or 45.5% of the national total.[5] The deployment of new infrastructure has accelerated. The world's largest Narrow Band Internet of Things (NB-IoT) network has been built in China. The enhanced Machine-Type Communication (eMTC) network is being deployed.

2. Reinforced Independent Innovation Capacity in Network Information Technology

General Secretary Xi Jinping stressed that, as a pillar of the nation, key and core technologies are of great significance to promoting high-quality economic development and protecting national security in China. He called on all of the country to make breakthroughs in the innovation of critical generic technologies, cutting-edge technologies, modern engineering technologies and disruptive technologies, and strive to make key and core technologies independent and controllable. China has thoroughly implemented the innovation-driven development strategy, strengthened the capacity construction of independent innovation in the Internet technology industry and accelerated the development of cutting-edge technologies such as AI, quantum computing and neural network chips. Breakthroughs have been made in fields such

---

[2]Data Source: The Ministry of Industry and Information Technology of the People's Republic of China (MIIT).

[3]Data Source: The 44th "Statistical Report on China Internet Development" released by China Internet Network Information Center (CNNIC), August 30, 2019, see http://www.cnnic.net.cn/hlw fzyj/hlwxzbg/hlwtjbg/201908/t20190830_70800.htm.

[4]Data Source: *The White Paper on the Development of IPv6 in China* issued by the Expert Committee for the Large-scale Deployment of IPv6.

[5]Data Source: The Ministry of Industry and Information Technology of the People's Republic of China (MIIT).

as high-performance computing, software technology, integrated circuit technology, cloud computing and big data. Thanks to the continuous growth of R&D investment, breakthroughs have also been made in areas such as IoT OS and Application-Specific Integrated Circuit (ASIC) chips, and the prototype of China's new-generation exascale supercomputer has been developed. Significant progress has been made in the development and application of cutting-edge technologies and asymmetric technologies in the field of network information. The 7 nm process technology has been deployed in AI chips. A large number of intelligent hardware computing platforms have been developed. The quantum key distribution protocol has reached the international leading level. Edge computing technologies have been gradually applied and an ecosystem of edge computing has been established. A new opportunity of accelerating the integration of Virtual Reality (VR) and traditional industries has emerged. In 2018, the size cloud computing in China hit ¥ 96.3 billion, up 39.2% over the previous year[6]; the size of China's big data industry reached ¥ 540.5 billion, up 14% over the previous year. Major Internet enterprises, information device manufacturers, and telecom operators have become an important force of IT research and development, which has accelerated the application of cutting-edge technologies such as AI, IoT, edge computing, and VR.

3. Thriving Digital Economy

General Secretary Xi Jinping stressed that the cyberspace affairs represent new productive forces and a new development direction, and that China should also be able to take a step forward in practicing new development ideas. It should promote the deep integration of the Internet, big data, and AI with real economy, develop and strengthen digital economy, accelerate the progress of digital industrialization[7] and industrial digitalization,[8] and give play to the amplification, superposition, and multiplying role of digitalization in economic development. With the deep integration of Internet and economy, e-commerce and Internet information service industries are booming, and new business models and forms for the integrated development of the Internet and industries are constantly emerging, which have provided a new impetus

---

[6]Data Source: *The White Paper on the Development of Cloud Computing (2019)* issued by China Academy of Information and Communications Technology.

[7]Digital industrialization refers to the information and communications industry, which includes the electronics and information manufacturing industry, telecommunications industry, software and information technology service industry, and the Internet industry. The definition is sourced from *The White Paper on Development and Employment of China's Digital Economy* released by China Academy of Information and Communications Technology.

[8]Industrial digitalization means the application of digital technologies into traditional industries, which results in the growth of both production quantity and production efficiency. And the additional outputs resulting therefrom become an important component of digital economy. The definition is sourced from *The White Paper on Development and Employment of China's Digital Economy* released by China Academy of Information and Communications Technology.

for the structural optimization of economic development and the transition from old to new drivers of growth. China's digital economy reached ¥ 31.3 trillion in 2018, accounting for 34.8% of the country's total GDP.[9] Digital economy has become a new engine of China's economic growth. Structurally, the value of digital industrialization in 2018 amounted to ¥ 6.4 trillion, entering a period of steady growth, and the value of industrial digitization grew rapidly to reach ¥ 24.9 trillion.[10] The integration of digital economy and real economy has continuously deepened. E-commerce is booming. In 2018, China's e-commerce trade volume reached ¥ 31.6 trillion in 2018, an increase of 8.5% over 2017; and the operating income of e-commerce industry reached ¥ 3.5 trillion, an increase of 20.3% over 2017.[11] *E-Commerce Law of the People's Republic of China* was officially enacted at the beginning of 2019. The development of China's e-commerce industry has ushered in a standardization stage. Laws have been put in place for online shopping, online payment, etc. Market activities are more standardized and consumers' rights and interests are now better protected. In the first half of 2019, China's online retail sales totaled ¥ 4.82 trillion, with an increase of 17.8% year on year.[12] The booming digital economy has spawned a large number of new business forms and new occupations and has become a new channel for optimizing employment structure and realizing stable employment. New models of digital economy, such as livestreaming and sharing economy, have spurred a sharp increase in the number of flexible employees. In 2018, some 191 million jobs were offered in digital economy, accounting for 24.6% of the overall employment.[13]

4. More Actions for People to Benefit More from the Internet and Informatization

General Secretary Xi Jinping stressed that cyberspace affairs must uphold a people-centered development vision, with people's well-being as the starting point and footing for informatization development, and must aim to enhance people's sense of gain, happiness, and security through informatization development. As of June 2019, the number of Internet users in China hit 854 million, with the Internet penetration rate reaching 61.2% and the number of websites reaching 5.18 million.[14]

---

[9]Data Source: *The Digital China Construction and Development Report (2018)* released by the State Internet Information Office.

[10]Data Source: China Academy of Information and Communications Technology.

[11]Data Source: *The China E-Commerce Report (2018)* released by the Ministry of Commerce.

[12]Data Source: The 44th "Statistical Report on China Internet Development" released by China Internet Network Information Center (CNNIC), August 30, 2019,
   see http://www.cnnic.net.cn/hlwfzyj/hlwxzbg/hlwtjbg/201908/t20190830_70800.htm.

[13]Data Source: *The White Paper on Development and Employment of China's Digital Economy* released by China Academy of Information and Communications Technology.

[14]Data Source: The 44th "Statistical Report on China Internet Development" released by China Internet Network Information Center (CNNIC), August 30, 2019, see http://www.cnnic.net.cn/hlw fzyj/hlwxzbg/hlwtjbg/201908/t20190830_70800.htm.

As network applications further thrive, which include livestreaming, online music and online education, high-quality and personalized contents are emerging, new entertainment forms such as short videos and vlogs are constantly launched, and high-quality educational and cultural resources are now available to an increasing number of people. As of June 2019, the number of users of livestreaming, online music, online video, and other network applications has increased by more than 30 million in half a year, respectively, and the number of online educational programme subscribers has reached 232 million, with a semi-annual increase of 15.5%.[15] Chinese people's educational, cultural, and entertainment demands have been greatly satisfied. China has attached great importance to utilizing information tools to promote the development of e-government services. Through the active launch of the National Integrated Online Government Service Platform, Chinese people's right to know and satisfaction have been greatly enhanced. As of June 2019, the number of e-government service users in China reached 509 million, or 59.6% of all Internet users. All the 31 provinces (autonomous regions, the municipalities directly under the Central Government), Xinjiang Production and Construction Corps, and more than 40 departments of the State Council have all registered on the National Integrated Online Government Service Platform. They have worked to enable people to access more government services via a one-stop service.[16]

5. Enhanced Construction and Management of Online Contents

General Secretary Xi Jinping stressed that China's Internet development should follow the principle of "maintaining positive energy, keeping things under control and correctly utilizing the Internet", accurately grasp the rule of the generation and evolution of online public opinions, constantly bring forth innovation in working philosophy, methods, carriers and channels, and institutional mechanisms, and improve the level of network governance, so that the largest variable of the Internet becomes the largest increase in career development. China has continuously strengthened the construction and management of online contents, actively spread positive energy, and thoroughly publicized Xi Jinping Thought on Socialism with Chinese Characteristics for a New Era, the spirit of the 19th National Congress of the CPC, as well as the great achievements of the past 70 years since the founding of the People's Republic of China, especially the new achievement and new progress made after the 18th National Congress of the CPC. China has continued to carefully plan major

---

[15]Data Source: The 44th "Statistical Report on China Internet Development" released by China Internet Network Information Center (CNNIC), August 30, 2019, see http://www.cnnic.net.cn/hlw fzyj/hlwxzbg/hlwtjbg/201908/t20190830_70800.htm.

[16]Data Source: The 44th "Statistical Report on China Internet Development" released by China Internet Network Information Center (CNNIC), August 30, 2019, see http://www.cnnic.net.cn/hlw fzyj/hlwxzbg/hlwtjbg/201908/t20190830_70800.htm.

online publicity campaigns, deepened education on the socialism with Chinese characteristics and the Chinese Dream, cultivated and practiced core socialist values, fostered healthy and positive Internet culture, and nourished people and society with mainstream ideological values and moral culture. It has vigorously pressed ahead with innovation in the concepts, contents, forms, methods, and approaches of online publicity while deeply analyzing online contents, focusing on improving the online experience of Internet users, and constantly improving the communicability, guiding power, influence and credibility of news media. It has formulated *Opinions on Speeding up the Establishment of a Comprehensive Internet Governance System*, and promoted the establishment of a comprehensive Internet governance system that provides integrated functions such as leadership management, positive energy dissemination, content management, social synergy, rule of law in cyberspace, and the use of technology in cyber governance, with an aim to improve China's comprehensive cyber governance capacity in an all-round way. A pilot anti-addiction system targeting teenagers has been launched and constantly expanded, which has effectively guided them to utilize the Internet healthily. Special campaigns to crack down cyber piracy and copyright infringement have also been actively implemented, including "Jian Wang 2019", "Jing Wang 2019", "Hu Miao 2019", and "Qiu Feng 2019". The mechanism for improving illegal information and website linkage and the whistle-blowing mechanism have been improved. For the first half of 2019, China's Internet complaint departments at all levels received a total of 68.579 million complaints, a year-on-year growth of 8.9%,[17] indicating that great efforts have been made to create a clean cyberspace.

6. Steadily Improved Ability to Protect Cybersecurity

General Secretary Xi Jinping stressed that without cybersecurity, there would be no national security, and no economic and social stability, nor interests of the broad masses of people be ensured. Facing the severe and complicated cybersecurity situation, China has strengthened the construction of cybersecurity protection system, built all-dimensional defenses for cybersecurity, improved its ability and level in protecting cybersecurity, and effectively responded to and resolved cybersecurity threats. It has taken comprehensive measures to protect key information infrastructure, established a framework for security protection of critical information infrastructure, and further carried out extensive cybersecurity inspections. It has strengthened data security administration, improved data security measures and advanced solidly big data security projects. It has enhanced the protection of personal information, deeply carried out the special campaign against the collection and use of personal information by apps in violation of laws and regulations, and actively regulated the

---

[17]Data Source: The 44th "Statistical Report on China Internet Development" released by China Internet Network Information Center (CNNIC), August 30, 2019, see http://www.cnnic.net.cn/hlw fzyj/hlwxzbg/hlwtjbg/201908/t20190830_70800.htm.

behaviors of personal data collection. Efforts have also been made in strengthening vulnerability management for information security. In the first half of 2019, China's National Information Security Vulnerability Sharing Platform reported 5,859 generic security bugs, a decrease of 24.4% over the same period of the previous year, among which there were 2,055 high-risk vulnerabilities, a decrease of 21.2% over the same period of the previous year.[18] China has accelerated the progress in issuing supportive regulations for *Cybersecurity Law* and has issued *Provisions on the Cyber Protection of Children's Personal Information*, China's first legislation on the cyber protection for children.

7. Deepened International Exchange and Cooperation on Cyberspace

General Secretary Xi Jinping stressed that all countries should deepen pragmatic cooperation with mutual improvement as the driving force and mutual win as the target, to explore a new path of mutual trust and collective governance and inject more vigor and vitality into the community of shared future in cyberspace. China has taken the "Four Principles" and the "Five Propositions" proposed by General Secretary Xi Jinping as the guide, comprehensively strengthened international exchanges and cooperation on cyberspace, and provided the China Program for global Internet development and governance. It has successfully hosted World Internet Conference for five consecutive years during which it has expanded brand effect, established cooperation platforms, and signed a series of cooperation agreements with governments, social organizations and enterprises in other countries. It has strengthened conceptual interpretation by publishing a series of concept papers on the "community of shared future in cyberspace", and furthered promoted cooperation and exchanges on cyberspace. It has been deeply involved in multilateral activities of global cyberspace governance through UN Internet Governance Forum (IGF), Internet Corporation for Assigned Names and Numbers (ICANN), World Economic Forum (WEF), and other platforms. It has consolidated and promoted exchanges and cooperation on cyberspace, including proceeding with its bilateral exchanges and cooperation with the United States, Russia, EU and other countries, and deepening cooperation and exchanges with emerging markets and other developing countries in cyberspace affairs. It has furthered cooperation with countries along the "Belt and Road" in digital economy and informatization construction, and promoted the implementation of the "Belt and Road" Digital Economy International Cooperation Initiative.

---

[18]Data Source: The 44th "Statistical Report on China Internet Development" released by China Internet Network Information Center (CNNIC), August 30, 2019,

   see http://www.cnnic.net.cn/hlwfzyj/hlwxzbg/hlwtjbg/201908/t20190830_70800.htm.

II. Innovation and Achievement of Internet Development in China's Provinces (Autonomous Regions, Municipalities)

China Internet Development Report has been publishing the China's Internet Development Index (CIDI) since 2017, which remains the same this year. Guided by General Secretary Xi Jinping's important thought on building China's strength in cyberspace, CIDI aims to comprehensively assess Internet development results and levels in China's 31 provinces (autonomous regions, municipalities directly under the Central Government, excluding Hong Kong, Macao, and Taiwan) by constructing an objective, authentic and accurate comprehensive assessing index system. On this basis, it attempts to guide these regions to further clarify their strategic goals and priorities of Internet development, accurately analyze their own comparative advantages, regional advantages and development advantages, and promote cyberspace affairs toward the goal of building China's strength in cyberspace.

CIDI is the weighted result of six indexes, namely, information infrastructure construction, innovation capacity, digital economy development, Internet application, cybersecurity and Internet governance, which gives a panorama of Internet development status in China's 31 provinces (autonomous regions, municipalities directly under the Central Government), and provides a quantitative basis for their Internet development. The 2019 assessment index is updated and improved on the basis of the 2018 index system by comprehensively taking into account Internet development in different regions and fully drawing on the opinions from relevant national departments, provinces and cities, the Cyberspace Affairs Think Tank and experts in related fields. First, the first-level indicators remain unchanged, while the second-level indicators are adjusted slightly. A comprehensive assessment system combining aggregate indicators and per capita indicators, and quantitative and qualitative indicators has been designed. Second, the weights of indicators are re-designed. They are determined by using the analytic method of hierarchical weight decision-making through questionnaires and solicitation of industry professionals' evaluation of indicators, which has improved the authoritativeness, scientificity and accuracy of the index system. Specifically, the weight of information infrastructure construction is changed from 10% to 18%, that of innovation capacity is changed from 20% to 18%, that of digital economy development is changed from 20% to 19%, that of Internet application is changed from 25% to 18%, that of cybersecurity is changed from 13% to 15%, and that of Internet governance remains unchanged (12%). See Table. 1.

In order to ensure the authenticity, integrity and accuracy of data, the evaluation data of 2019 China Internet Development Index mainly has two sources: first, the statistical data and index from national departments such as the Office of Central Cyberspace Affairs Commission, National Bureau of Statistics, Ministry of Industry and Information Technology (MIIT), Ministry of Science and Technology (MOST), and CNNIC; second, relevant statistical data from cyberspace affairs departments of each province (autonomous regions, municipalities directly under the Central Government).

Table 1 China's Internet Development Index System

First-level indicators	Weight	Second-level indicators	Description
Information infrastructure construction	18%	Broadband infrastructure	Per capita Internet broadband access ports, proportion of fiber-optic broadband subscribers, broadband network rate, etc.
		Mobile infrastructure	4G network download speed, proportion of 4G mobile phone subscribers, etc.
Innovation capacity	18%	Innovation environment	Total number of incubators, number of R&D personnel, per capita GDP, etc.
		Innovation input	Proportion of local government spending on science and technology, proportion of R&D expenditure to GDP, proportion of corporate R&D personnel, etc.
		Innovation output	Number of scientific papers per 10,000 heads, number of national awards for scientific and technological achievements, number of invention patents per 10,000 heads, etc.
Digital economy development	19%	Basic indicator	Number of Internet users, business volume of telecommunication services, proportion of the added value of information transmission and information technology services, etc.
		Integration indicator	Intelligent manufacturing readiness rate, numerical control rate of key processes, digitization rate of production equipment, etc.
		Industrial indicator	Proportion of e-commerce consumption expenditure in consumption expenditure, development of the big data industry and AI technology industry, number of China's top 100 Internet companies, number of unicorn companies, etc.
Internet applications	18%	Personal application	Penetration rate of mobile phone, usage rate of social applications, audio-visual applications and daily life service applications, etc.
		Enterprise application	Application rate of industrial cloud platforms, proportion of networked collaboration enterprises, proportion of service-oriented manufacturing enterprises, proportion of enterprises engaging in e-commerce transactions, etc.
		Public application	E-government service ability of provincial governments, number of network multimedia classrooms per 100 students, etc.
Cybersecurity	15%	Cybersecurity environment	Number of controlled malicious computer programs, proportion of infection hosts in the number of local active IP addresses, IP address distribution of devices controlled by IOT malicious codes, etc.
		Cybersecurity awareness construction	Cybersecurity search index, cybersecurity information index, page views of cybersecurity we-media.

(continued)

Table 1 (continued)

First-level indicators	Weight	Second-level indicators	Description
		Cybersecurity industrial development	Number of cybersecurity enterprises, number of top 100 cybersecurity companies, etc.
Internet governance	12%	Construction of network management institutions	Setting of provincial cyberspace affairs departments, number of cyber societies, etc.
		Construction of media convergence	Number of verified Weibo accounts of government departments, WeChat communication index on provincial government affairs, number of government Toutiao accounts, etc.
		Construction of Internet governance system	Number of issued provincial and municipal regulations, policies and action plans, number of Internet information licenses, etc.

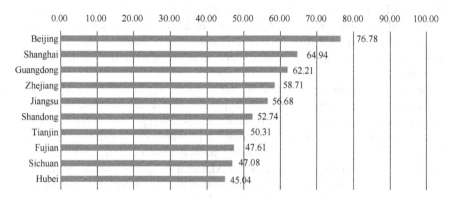

Fig. 1 List of Top 10 Provinces (Autonomous Regions, Municipalities Directly Under the Central Government) in the Composite Ranking of China's Internet Development Index in 2019

1. CIDI 2019 Composite Ranking

Based on the China Internet development indicator system, Fig. 1 provides the list of top 10 provinces (autonomous regions, municipalities directly under the Central Government) in the composite ranking of China's Internet development index in 2019. As shown in the Fig. 1, economically developed regions such as Beijing, Shanghai, Guangdong, and Zhejiang are among the top regions in Internet development, and the central and western regions also increase their effort and display a strong momentum of growth.

Taken together, China's provinces, autonomous regions and municipalities directly under the Central Government have deeply implemented General Secretary Xi Jinping's important thought on building China's strength in cyberspace and earnestly acted on the central government's relevant requirements. They have intensified web content construction and governance, cybersecurity protection, informatization development and other relevant work, promoted innovation in development thoughts, and carried out a series of distinctive innovative practices based on local conditions. As a result, significant achievements have been made in China's Internet development. In the composite rankings of China's Internet development index in 2019, Beijing, Shanghai, Guangdong, Zhejiang, Jiangsu, Shandong, Tianjin, Fujian, Sichuan, and Hubei are the top ten regions in Internet development.

Beijing has vigorously intensified efforts in information infrastructure construction, strengthened industrial integration and increased innovation input. It has accelerated the development of big data and AI, fully promoted the integrated application of big data, IoT, cloud computing and other technologies, and attached importance to cyber environment construction. Beijing is currently home to the largest number of cybersecurity enterprises in China.

Shanghai has fully advanced intelligent city construction, accelerated technological R&D and the development of innovative industries, and steadily developed Internet information services and e-commerce. Besides, it has comprehensively accelerated international development, and continued to improve Internet convenience services, leading to a gradual increase in the market size of the Internet industry.

Guangdong has taken the lead in 5G development, vigorously promoted "5G+" application and development, greatly expanded the scale of software industry (ranking the first in the country), carried out demonstration projects of big data application, vigorously promoted the development of manufacturing industries, established AI industrial clusters, continuously pushed forward entrepreneurship and development in emerging industries such as AI, facial recognition, unmanned driving and smart health care, and sped up the construction of "Internet + government services" by launching the applet "Yueshengshi".

Zhejiang is the first region in China to realize full coverage of 4G network and fiber optic network. It has rapidly developed the basic network information technology, made breakthroughs in cloud computing, big data, IoT and AI by developing new intelligent city applications, deeply implemented digital economy "No. 1 Project", and led the high-quality development of industrial economy. Besides, it has continued to improve the construction of digital information platforms, made great achievements in digital government construction, further implemented the "One Visit at Most" reform, and led the whole nation in the field of online inclusive education.

Jiangsu has actively promoted technological R&D and innovation. As of June 2019, the province had a total of 50,367 information technology patents. It had rapidly developed e-commerce industry and deeply promoted the "One Village, One Product, One Online Store" Project. With distinct advantages in IoT industry, it has undertaken or participated in Internet application projects in more than 20 countries around the world. The province ranks top in the convergence of informatization and industrialization in the whole nation, and promotes the network construction of government services in accordance with high standards.

Shandong has optimized its information industrial ecosystem, established major scientific and technological innovation platforms, strengthened network environment construction, accelerated the convergence of high-level Internet talents, and promoted the training of Internet talents. By rapidly developing intelligent manufacturing, Shandong ranks top in the number of intelligent manufacturing pilot projects in China. In addition, it has showed a good momentum of development in industrial Internet platforms, and explored a suitable path of integrated development of manufacturing and the Internet.

Tianjin has actively encouraged Internet enterprises to continually engage in product innovation and technological innovation, accelerated the aggregation of innovative elements in enterprises, and promoted the establishment of a management system for the convergence of informatization and industrialization. The city has focused on developing key areas such as R&D, design, production, management and marketing, deepened the integration of manufacturing and Internet, and fully supported intelligent manufacturing enterprises. A multi-dimensional and gradient cultivation mechanism for digitalization, informatization, and intelligentization has thus taken shape.

Fujian has actively carried out the leading action of developing digital economy, and vigorously implemented the pilot projects for the integration and innovation of the Internet and industry. The province has a total of 323 enterprises passing the evaluation of the informatization and industrialization convergence management system, ranking top in the whole nation. It has taken the lead in the construction of national healthcare big data platforms and security service platforms in the country, and made remarkable achievements in cybersecurity governance.

Sichuan has comprehensively promoted the construction of Gigabit Optical Network, steadily pushed forward the development of e-commerce industry and the software and information service industry, accelerated the construction of government network infrastructure, actively built innovation platforms for industrial Internet, accelerated the transformation of industrial digitalization, and made positive progress in the convergence of informatization and industrialization.

Hubei has sped up the construction of industrial Internet platforms, improved the industrial ecosystem and attached importance to strengthening its security assurance ability. It has taken the lead in IPv6 construction in the country, focused on "making breakthroughs in core technologies" such as chips, 5G, new display technology and optical communication, and effectively enhanced the environment and basic conditions for "Internet + industry" development. As a result, new business forms and models have been constantly emerging and traditional enterprises have been upgraded.

Table 2 List of Top 10 Provinces (Autonomous Regions, Municipalities Directly Under the Central Government) in the Ranking of Single-item Assessment Indexes

Ranking	Information infrastructure construction index	Innovation capacity index	Digital economy development index	Internet applications index	Cybersecurity index	Internet governance index
1	Beijing	Beijing	Beijing	Beijing	Guangdong	Shandong
2	Shanghai	Shanghai	Shanghai	Zhejiang	Beijing	Hebei
3	Jiangsu	Tianjin	Guangdong	Shanghai	Shanghai	Tibet
4	Zhejiang	Jiangsu	Zhejiang	Jiangsu	Fujian	Beijing
5	Fujian	Guangdong	Jiangsu	Guangdong	Sichuan	Jiangsu
6	Guangdong	Zhejiang	Shandong	Shandong	Jiangsu	Zhejiang
7	Liaoning	Anhui	Sichuan	Sichuan	Zhejiang	Guangdong
8	Tianjin	Hubei	Tianjin	Guizhou	Hubei	Henan
9	Shandong	Shandong	Fujian	Chongqing	Tianjin	Sichuan
10	Ningxia	Shaanxi	Chongqing	Fujian	Chongqing	Inner Mongolia

2. Itemized Score Index and Rankings

Table 2 presents the list of top 10 provinces (autonomous regions, municipalities directly under the Central Government) in the ranking of 6 areas, namely information infrastructure construction, innovation capacity, development of digital economy, Internet application, cybersecurity and Internet governance, based on the China Internet development indicator system.

3. Rankings of Information Infrastructure Construction Index

All regions in China have attached great importance to information infrastructure construction. They have focused on optimizing and improving network quality and actively promoted pilot commercial 5G network deployment and IPv6 scale deployment. With these efforts, a system of broadband and mobile network infrastructure and services that enables high-speed and unblocked access to the Internet, covers both urban and rural areas and provides convenient services has been established. The top ten regions in information infrastructure construction are Beijing, Shanghai, Jiangsu, Zhejiang, Fujian, Guangdong, Liaoning, Tianjin, Shandong, and Ningxia. In 2018, Beijing had the largest number of per capita Internet broadband access ports in China, reaching 96.92 per 100 people.[19] Shanghai ranks first in the average download rate of fixed broadband and 4G network. Jiangsu has accelerated the construction of new-generation information infrastructure and increased investment in information

---

[19] Source: Beijing Cyberspace Administration.

infrastructure. The total length of optical cable lines in the province is up to 3.53 million kilometers, ranking first in China. All the administrative villages in Jiangsu Province are connected to the optical network.[20] Zhejiang is the first in China to complete the construction of "Optical Network City" and its number of households with Fibre-to-the-Premises (FTTP) connections is among the top in China. Besides, its backbone network capacity has been comprehensively upgraded.

4. Rankings of Innovation Capacity Index

All regions in China have constantly provided financial and personnel support for the R&D of innovative technologies, actively promoted innovative research and break-throughs in new-generation information technologies represented by AI and big data, and made new progress in cutting-edge technologies. Among them, the top ten regions in innovation capacity are Beijing, Shanghai, Tianjin, Jiangsu, Guangdong, Zhejiang, Anhui, Hubei, Shandong and Shaanxi. Beijing has become a high and new tech innovation hub in China and ranks first in a variety of indicators, including innovation input, innovation environment and innovation output. Tianjin has stepped up efforts in developing new virtualization technology, critical AI technologies and core blockchain technologies, vigorously promoted the integration of manufacturing industry and the Internet, and utilized its strengths to build "Tianjin Smart Port". Shaanxi has systematically promoted the construction of industrial Internet infrastructure and the data resource management system, as well as the building of a local cluster of cyberspace enterprises with capacity for independent innovation, many of which have certain competitive advantages and influence in the industry.

5. Rankings of Digital Economy Development Index

All regions in China have attached great importance to digital economy development. They have issued relevant plans to promote the development of digital economy and digital industry, accelerated the progress of digital industrialization and industrial digitization, and produced remarkable results in the informatization of manufacturing industry. Among them, the top ten regions in digital economy development are Beijing, Shanghai, Guangdong, Zhejiang, Jiangsu, Shandong, Sichuan, Tianjin, Fujian and Chongqing, with their development levels all above the national average. In 2018, Beijing had the largest number of "Internet Top 100 Enterprises" and "unicorn companies" in China, reaching 32[21] and 87,[22] respectively. In Shanghai, the e-commerce industry has developed rapidly and robustly and its market size has

---

[20] Source: Jiangsu Cyberspace Administration.

[21] Data Source: *China's Top 100 Internet Companies in 2018* released by the Internet Society of China.

[22] Data Source: *Research Report on China's Unicorn Companies in 2018* issued by Prospective Industry Research Institute.

stayed ahead in China. Guangdong's software industry has topped one trillion *yuan* for the first time, ranking first in China. In 2018, the province's software export volume reached 26.73 billion US dollars, accounting for 52.4% of the country's total.[23] Chongqing has deeply implemented the action plan for the innovation-driven development strategy with big data intelligentization playing a leading role, and the intelligent industries have maintained a sound momentum of development.

6. Rankings of Internet Application Index

All regions in China have attached great importance to the Internet industry and its deep integration with other industries. Remarkable achievements have been made consequently, including the rapid growth of the Internet industry, continuous integration of the Internet and industry, and deep integration of the Internet and public services in fields such as government affairs and education. The top ten regions in Internet applications are Beijing, Zhejiang, Shanghai, Jiangsu, Guangdong, Shandong, Sichuan, Guizhou, Chongqing, and Fujian. For personal Internet applications, by the end of 2018, Beijing had the highest mobile phone penetration rate and its usage rate of Internet social applications, audio-visual applications and daily life service applications all ranked forefront in the country. Zhejiang has actively promoted the digital construction of manufacturing industry and continuously deepened the digital transformation of the service industry. It ranks first in the index of Internet enterprise applications. Guangdong has actively developed informatization and e-government services and taken the lead to launch "Yueshengshi", an applet of livelihood services. It ranks first in the capacity of e-government services in the country. Fujian has vigorously promoted smart city construction through the "Digital Fujian" project and continuously enhanced its intelligence level of urban governance.

7. Rankings of Cybersecurity Index

All regions in China have attached great importance to cybersecurity work. They have actively promoted the cybersecurity protection capacity building, continued to enhance their capacity to protect cybersecurity, and constantly improved the safety emergency response system. Besides, a rich variety of cybersecurity publicity campaigns have been carried out to promote the rapid development of the cybersecurity industry. The top ten regions in cybersecurity are Guangdong, Beijing, Shanghai, Fujian, Sichuan, Jiangsu, Zhejiang, Hubei, Tianjin, and Chongqing. The cybersecurity industry has been flourishing in Guangdong, Shanghai and Fujian, which have the largest number of cybersecurity enterprises in China. Sichuan has stepped up efforts in cybersecurity personnel training and fully integrated various resources, including scientific research institutions, universities, associations, and enterprises to explore a new talent training model for information industry. It has also promoted the

---

[23] Data Source: Guangdong Cyberspace Administration.

establishment of cyberspace security colleges within the province and strengthened the research on relevant technological policies and standards of information security. Hubei has actively promoted cybersecurity personnel training and industrial development, established a high-standard national base for cybersecurity personnel and innovation, and created a "creative valley" model incorporating first-class cybersecurity colleges and first-class cybersecurity industries with Chinese characteristics.

8. Rankings of Internet Governance Index

All regions in China have attached great importance to Internet governance. They have actively issued relevant regulations, industrial policies and action plans for Internet governance, strengthened the top-level design of Internet governance, constantly improved the functions of cyberspace institutions, rapidly developed Internet societies, and continued to enhance the construction level of media convergence. The top ten regions in Internet governance are Shandong, Hebei, Tibet, Beijing, Jiangsu, Zhejiang, Guangdong, Henan, Sichuan and Inner Mongolia. Shandong has constantly improved the top-level design of Internet governance and actively promoted the implementation of the national cyber development strategy and the Digital China strategy. It has issued *Digital Shandong Development Plan (2018–2022)*, aiming to build an innovative digital governance pattern for the development of digital Shandong. In addition, it has actively improved supporting policies, promoted the formulation of informatization policy documents and standards, and strengthened the construction of government service platforms to standardize the management of the new media of government affairs. Hebei has improved the system for cyberspace affairs management, accelerated the construction of cyberspace affairs institutions and formulated a series of policy documents to promote law-based cyberspace governance. Besides, through the active construction of media convergence and Internet societies, an extensive, collaborative and efficient mobile communication system has been established. Tibet has attached great importance to the construction of cyberspace affairs institutions. It has constantly improved the comprehensive Internet governance system and its Internet ecology has been improved on the whole. Beijing has further deepened Internet governance and actively issued important policy documents on cyberspace affairs. Jiangsu has constantly strengthened the construction of Internet societies where the number of Internet societies is among the highest in China. It has actively given play to the role of Internet societies in Internet industrial self-discipline, comprehensive governance of network ecosystem, Internet publicity, Internet-based charity, Internet-assisted poverty alleviation, cybersecurity publicity, among others.

III. Prospects of China's Internet Development

As socialism with Chinese characteristics has crossed the threshold into a new era, China's Internet development has also entered an important period of strategic opportunities. Having reached a new historical starting point, it is imperative that China's Internet development adheres to the guidance of Xi Jinping Thought on Socialism with Chinese Characteristics for a New Era, particularly his thought on building China's strength in cyberspace, focuses on realizing the historic task of the great rejuvenation of the Chinese nation, follows the trend of information revolution age, implements the historic mission of building China's strength in cyberspace, brings more benefits to the country and its people, and makes positive contributions to realizing the Two Centenary goals and the great rejuvenation of the Chinese nation.

1. Information Infrastructure Construction Will Become a New Fulcrum to Support Development and Transformation

Infrastructure is a critical pillar of China's economic and social development and an important channel of logistics, personnel flow, capital flow and information flow. With the rapid development and popularization of new-generation information technologies and applications such as cloud computing, big data, AI and blockchain, information infrastructure is playing an increasingly strategic, fundamental, and leading role in economic and social transformation and development. In particular, new-generation infrastructure represented by 5G, AI, IoT, industrial Internet, and satellite Internet is deeply integrated with economic and social fields, and is a strategic basis and important pillar of the current digital transformation of China's economy and society. At present, new-generation infrastructure system suited to digital economy and modern governance is still in need of improvement, China's capacity of providing information services is inadequate, the urban-rural and cross-regional digital divide is not completely eliminated, and the mechanism for the co-constructing and of data resources is imperfect yet. Facing the future, China should speed up the deployment of new-generation information infrastructure, promote the construction of intelligent information infrastructure, actively press ahead with 5G network's scale deployment, prioritize information infrastructure as a key field of investment and accelerate the construction of new-generation fast, mobile, secure and ubiquitous information infrastructure. It should continue to improve the quality and efficiency of broadband network, further enhance the coverage and quality of 4G network, and encourage enterprises to carry out pilot and demonstration projects of innovative applications such as 5G, IPv6, IoT and AI, and popularize intelligent cities, Internet of Vehicles (IoV), intelligent health care and other applications and services. It should vigorously promote the coordinated development of regional infrastructure and extension of informatization to underlying infrastructure, focus on developing informatization of grassroots services, step up investment in network construction in central and western regions and rural areas, constantly correct regional development imbalance, and implement the digital rural development strategy for urban and rural grassroots organizations, especially the vast rural areas.

2. Digital Economy Will Continue Growing with Higher Quality, Contributing to Economic Transformation

Developing digital economy is of great and far-reaching significance for implementing the decisions and plans of the CPC Central Committee, deepening the supply-side structural reform, promoting the shift in driving forces for development, constructing the modern economic system, and achieving high-quality development. At present, facing the complicated and challenging international environment and downward economic pressure, China's digital economy maintains rapid development, and new technologies and new business forms and models emerge, with the structure of digital economy continuously optimized, steady progress achieved in digital industrialization, and industrial digitalization further deepened. For quite a long period to come, China will still be in a critical period of transition from rapid growth to high-quality development, during which the rapid penetration and diffusion of emerging technologies will provide new momentum for economic growth, and the rapid development of digital economy that takes digital information as the factor of product will create important opportunities for improving the digitalization, networking and intelligentization level of real economy and building China's global competitiveness. Facing the future, China should deepen digital industrialization and industrial digitalization, carry out extensive digital application and model innovation, give play to the amplification, superposition and multiplying role of digitalization in economic development, fully promote the deep integration of the Internet, big data and AI with real economy, and push ahead the transformation and upgrading of traditional industries such as industry, services and agriculture sectors through digitalization and intelligentization means. It should formulate relevant policies on digital economy and set an "observation period" for new forms of business that it has a limited knowledge of. At the same time, it should improve its ability of risk prevention, establish an early warning system for preventing relevant risks and crack down on illegal and criminal acts to ensure the sound and stable development of digital economy.

3. Cybersecurity Will Have to Tackle Interwoven Risks and Frequent Threats

China is currently confronted with severe cybersecurity circumstances, increasingly complicated cybersecurity environment and intensified cybersecurity risks.

Traditional cybersecurity threats should not be ignored. Vulnerabilities in basic software/hardware such as CPU chips pose a serious threat to cybersecurity. The frequency of Distributed Denial-of-Service (DDoS) attacks has decreased, but their peak traffic continues to increase. Advanced Persistent Threats (APTs) against national key industrial units occur frequently. The issue of large-scale leakage of user personal information remains severe. As new technologies and new applications such as mobile Internet, cloud platforms and networked intelligent devices are enriching people's digital life and the exposure of network is expanded, new threats and new risks are caused. Critical information infrastructure in industrial Internet, health care, electric power and other industries is facing more security issues. Despite the growing demand for cybersecurity of information infrastructure, service applications and data information, the protection system of critical information infrastructure is still in need of improvement, there are gaps in big data security management, and the supporting capacity of cybersecurity industries remains inadequate. Facing the future, China should foster correct cybersecurity concepts, deeply implement *Cybersecurity Law* and solidly advance the cybersecurity work. It should prioritize the work of protecting the safety of critical information infrastructure, fulfill its responsibilities for protecting critical information infrastructure, and improve the protection system of critical information infrastructure. It should strengthen data security administration, prioritize data security in the cybersecurity work, crack down on illegal and criminal acts, including hacking, telecommunications network fraud and invasion of citizens' rights of privacy, and safeguard people's legitimate rights and interests. It should increase investment in cybersecurity, expand the cybersecurity market, further optimize the development environment for cybersecurity, and promote the cybersecurity industry and cybersecurity enterprises to grow in strength. It should strengthen cybersecurity risk prevention, particularly reinforce the work on analyzing the potential risks in the development of AI, 5G and other fields, establish and improve relevant laws and regulations, institutional systems and moral principles, and ensure the safety, reliability and controllability of new technologies.

4. The Development of Internet Media Will Feature Changes, Innovation and Deep Convergence

At present, the new round of technological revolution represented by information technology has profoundly impacted the traditional communication pattern. New technologies, new applications and new forms of business represented by AI, 5G and AR are developing rapidly, which has greatly boosted the development of Internet media and promoted the intelligentization of content creation, distribution, dissemination, consumption and other links. As a result, content productivity is further improved. Relevant information is more precisely pushed and acquired. The attractiveness, expressiveness and appeal of media content are further enhanced. An intelligent media-dominated, IoT-based, intelligentization-oriented mobile communication system is taking shape, which will drive new changes in China's media landscape. At the same time, as the Internet becomes the main channel for producing, disseminating and obtaining information, the network's ability of social mobilization grows stronger and becomes a conductor and amplifier of various risks. Facing the future, China should adhere to the principle of "maintaining positive energy, keeping things under control and correctly utilizing the Internet", accurately grasp Internet communication rules, vigorously promote innovation in the concepts, working content, approaches and systems of network publicity, pay more attention to improving user experience, and deeply analyze relevant information. It should also be skilled at using new technologies to improve and innovate the network communication model, grasp the mobile, social and intelligent trend of network communication, and give full play to the characteristics and advantages of new technologies and new applications. It should utilize the achievements of information revolution to promote deep media convergence and accelerate building an integrated pattern of omni-media communication. It should improve Internet governance capacity, accelerate the establishment of a system of comprehensive Internet governance, and improve the laws and regulations system. At the same time, it should enhance its capacity of using technology in Internet governance, delegate the main responsibility to network platforms, and enhance the network content ecology. It should extensively mobilize and rely firmly on Internet users, make them producers, disseminators and leaders of positive energy, encourage them to influence and educate each other, and guide them to consciously regulate their online behaviors and purify cyberspace.

5. Patterns of Global Internet Governance Will Be Adjusted with Rules Developed

At present, the international environment of cyberspace undergoes profound changes, and international cyberspace governance enters an important transition period. New technologies and new applications such as AI, quantum computing, 5G and IoT are emerging and exerting an important influence on the Internet governance pattern. Disagreement remains over the international governance model. Emerging economies and developing countries are speeding up digital economy development, improving digital capabilities and gradually increasing their international power of discourse and influence, which has a new impact on the digital world. The existing international governance mechanism is difficult to adapt to the rapid development of the Internet and the situation of international governance, and the vulnerability and uncertainty of cyberspace, are further revealed. In recent years, China has actively engaged in international cyberspace governance, practically promoted international cooperation on cyberspace, and constantly driven international cyberspace governance towards a more equitable and reasonable direction. In the meantime, China is facing more severe and complex problems in international cyberspace governance.

Some countries even take information technology, products and services as an important means to attack and contain other countries, which aggravates the confrontational threats in cyberspace. It is an important mission for China to deeply participate in international cyberspace governance to give better play to its role as a responsible cyberpower, enhance its power of discourse and influence, and build a digital world of mutual trust and governance with other countries. Facing the future, China should promote international cooperation and exchange on cyberspace, focus on the Four Principles of the global Internet governance system transformation and the Five Propositions of building a cyberspace community of shared future, and promote the establishment of a more just and equitable global Internet governance system. It should actively participate in global Internet governance, deeply engage in the activities of important platforms for global cyberspace governance, actively promote the integration of digital economy and cybersecurity, continue to hold World Internet Conferences (WIC), and build an international platform of interconnection between China and the world, as well as a Chinese platform of Internet sharing and co-governance. It should adhere to a multilateral approach with multi-party participation, give full play to the role of various stakeholders, including governments, international organizations, Internet enterprises, tech communities, non-government institutions and individual citizens, and deepen international exchange and cooperation on cyberspace. It should take the opportunity of the "Belt and Road Initiative" (BRI) to strengthen cooperation with the countries along the Belt and Road on information infrastructure construction, digital economy, cybersecurity and other fields, and vigorously promote the construction of the 21st Century Digital Silk Road.

Facing the future, China's Internet development has broad prospects but a long way to go. We should warmly embrace the Internet, actively utilize it and vigorously develop it, making Internet development more beneficial to the country, the people and the world.

Contents

Chapter 1
Information Infrastructure Construction

1.1 Outline

As a strategic public infrastructure for economic and social development, information infrastructure is an important basis for developing new economy and fostering new growth drivers. Besides, it plays an important role in building China's strength in cyberspace and promoting its transformation and upgrading. With the rapid development of information technology, information-driven innovation is playing an increasingly leading role, which makes it urgent to accelerate the construction of information infrastructure and give full play to its strategic, fundamental and leading role in economic and social development. General Secretary Xi Jinping stresses that it is important to step up the development of information infrastructure, enhance the deep integration of information resources, and open up the information "artery" of economic and social development. The Central Economic Work Conference for 2018 proposed to "accelerate the commercial application of 5G and strengthen the construction of new infrastructure such as AI, industrial Internet and IoT". Over the past year, China's information infrastructure has been developing steadily; progress has been made in traditional broadband networks and application facilities; and new infrastructure has been expanding at an ever-increasing rate. As a result, China's information infrastructure is rapidly evolving into a new generation of information network infrastructure characterized with high speed, intelligence, ubiquity and security, which has effectively pushed forward China's economic and social development.

Progress has been made in traditional broadband networks and application facilities. China has the highest optical broadband penetration in the world, with many cities achieving full coverage of the "Gigabit Optical Network". It has an in-depth coverage of 4G network and the largest number of 4G users in the world. China ranks the second in the number of available IPv4 addresses and is witnessing a growth in the number of IPv6 addresses, with the notification rate increasing steadily. Besides, the New Generic Top-Level Domain (New gTLD) market begins to recover and the overall domain market returns to positive growth. With the acceleration of IPv6

© Publishing House of Electronics Industry 2021
Chinese Academy of Cyberspace Studies, *China Internet Development Report 2019*, https://doi.org/10.1007/978-981-33-6930-6_1

commercial deployment, telecom carriers' IPv6 network has generally stable performance. The number of data centers is increasing rapidly and their layout is gradually optimized, with large data centers having become the main force of growth.

The construction of new infrastructure such as 5G, AI, industrial Internet and IoT has been accelerating, and increasingly becoming the strategic basis for the transformation and development of economy and society. As the first year of China's 5G mobile broadband network commercialization is underway, the major links of industrial chain such as systems, chips and terminals have been commercially available. Progress has also been made in the construction of the Identification and Resolution System of Industrial Internet. Five national top-level nodes have been released and put into trial, and the top-level layout of "East, West, South, North and Center" has initially taken shape. Industrial Internet platforms have been deployed in some vertical market segments, and the industrial Internet safety system has made positive progress in establishing the security framework and standard system as well as the research and development of technical products. Remarkable results have been achieved in the construction of Narrow Band Internet of Things (NB-IoT), and eMTC network is currently being implemented. AI application is extended to the telecommunications domain, and the backbone network architecture keeps being optimized and will undergo intelligentized reconstruction. The content distribution network (CDN) industry has been developing continuously.

1.2   Progress in Broadband Network Construction

1.2.1   Optimized Backbone Network Structure

1.2.1.1   The Backbone Network Is Gradually Improved

The traffic grooming effect of Internet backbone straight points has shown, and the provinces with Internet backbone straight points have witnessed significant improvement in network performance. By the end of 2018, the interconnection bandwidth between Internet backbone straight points reached 6,800 GB/s, an increase of more than 300% over 2013. According to the monitoring data of the Internet monitoring and analysis platform of the China Academy of Information and Communications Technology (CAICT), the average network delay in the provinces with Internet backbone straight points was 39.59 ms by the end of 2018, with an average packet loss rate of 0.19%, which rose by 15.27% and 56.65% respectively compared with the end of 2017.

There has been an increasing demand for network interconnection, and new Internet exchange points have sped up the pace of exploration. Beyond backbone networks, with the rapid development of Internet, Internet enterprises, local Internet access providers, IDC enterprises and cloud service enterprises are undergoing fast development, generating a variety of demands for new traffic interaction. Industrial authorities are actively carrying out pilot projects of building new Internet exchange points to further improve efficiency and reduce cost of network interconnection.

1.2.1.2 Technological Innovation Is Boosting Network Intelligentization

As new technologies such as software-defined network/network function virtualization (SDN/NFV), big data and AI are gradually maturing, backbone networks are becoming more flexible and intelligent in order to host various types of business services. China's basic telecom operators actively explore intelligent control methods that are combined with SDN technologies to promote the flexible scheduling and rational, efficient utilization of network resources.

(1) SDN technologies are employed for the efficient traffic scheduling between data centers.
(2) SDN technologies are employed for the intelligent traffic scheduling of IP network. China's three major telecom operators, China Telecom, China Mobile, China Unicom, have explored to use SDN technologies for route optimization and load balancing of backbone networks and international Internet inlet and outlet traffic scheduling. SDN technologies have also been deployed in the existing network.
(3) SDN technologies are employed to coordinate IP network and optical transport network. China Mobile has actively explored to utilize SDN to efficiently coordinate resources of IP bearer network and optical transport network. Under such circumstance, it has realized unified control and management of multi-vendor and multi-domain networking environment.

1.2.1.3 Multi-layer Cloud-Network Coordination

Facing the trend of cloud-network coordination, China's basic telecom operators have built a network of data centers that are efficiently connected to basic network. China Telecom has deployed the Phase I project of building its third SDN network and actively promoted better cloud-network coordination through utilizing SDN technologies. China Unicom has incorporated 65 self-owned high-quality data centers into its data center network. It plans to realize full coverage of SDN network in all cities (prefectures) of China and further build more than 30 SDN support points by the end of the year. China Mobile plans to build a "4 + 45" network architecture of data center interconnection (DCI) and construct a new network for data traffic grooming across the country. Based on the development of cloud-network coordination, China's basic telecom operators have actively explored to build an intelligent and open bearer network, reconstruct the network architecture with data centers as the focus and provide cloud-network integration services.

1.2.2   Accelerated Deployment of Gigabit Optical Network

1.2.2.1   Solid Progress Is Made in the "Boosting Internet Speed and Lowering Internet Charges" Campaign

In 2018, the Ministry of Industry and Information Technology (MIIT) and the State-owned Assets Supervision and Administration Commission (SASAC) went on to carry out the "boosting Internet speed and lowering Internet charges" special campaign. Meanwhile, clear requirements were proposed in improving the network supply capacity, reducing broadband charges, popularizing high-speed broadband applications and optimizing the telecom market environment. As the campaign is further deepened, it has brought benefits to enterprises and the masses on an extensive scale.

In terms of "boosting Internet speed", basic telecom operators have constantly increased investment over the previous three years, which amounted to over 1.2 trillion *yuan*, to expand the coverage of fiber optic networks and forge ahead with fiber installation. The depth and breadth of 4G network coverage, including in-door spaces such as office and business buildings and elevators, has been increased. The quality of continuous coverage along railways and highways has been improved.

In terms of "lowering Internet charges", in order to benefit people to the greatest extent, China has solidly promoted the work of lowering the rates of home broadband, and corporate broadband and dedicated Internet access services, and comprehensively abolishing domestic data roaming charges by telecom operators. The rates for mobile Internet services have been cut by at least 30%.

1.2.2.2   The Construction of Gigabit Optical Network Is Accelerated

So far, China has achieved full coverage of fiber broadband network at the city level. More than 98% of administrative villages have been covered with optical fiber. In some regions, the fiber optic network has been expanded to natural villages. As of the end of June 2019, China Telecom, China Mobile and China Unicom had built a total of 810 million FTTH/O ports, accounting for 90% of all broadband ports. 91% of broadband users in China were on fiber optic, ranking the first in the world. While promoting the popularization of fiber optic network, telecom operators have accelerated the construction of Gigabit Optical Network and continuously improved the network supply capacity. In 2018, Shanghai achieved full coverage of the Gigabit Optical Network, and other cities such as Beijing, Hangzhou, Xi'an, Suzhou, Ningbo and Luoyang also initiated "Gigabit City" construction successively. Figure 1.1 shows the total number and penetration rate of fixed broadband optical fiber ports in China from 2013 to June 2019.

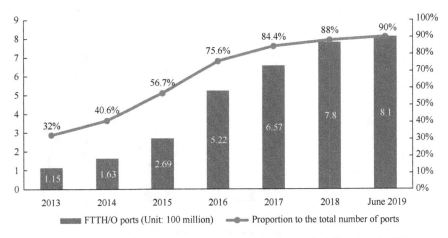

Fig. 1.1 Total number and penetration rate of fixed broadband optical fiber ports in China from 2013 to June 2019. *Data Source* The Ministry of Industry and Information Technology of the People's Republic of China (MIIT)

1.2.2.3 The Number of 100 Mbps Broadband Subscribers Has Been Increasing

The deployment of fiber optic network provides the essential condition for improving the rate of access to Internet by broadband users in China. As of the end of June 2019, the number of users with access rate at or above 100Mb/s reached 335 million, accounting for 77.1% of the total broadband users. And the proportion of users using high-bandwidth products has been increasing, as shown in Fig. 1.2.

1.2.3 China Entering the First Year for Commercial Uses of 5G Mobile Broadband Network

1.2.3.1 China Has Started the Deployment for Commercial Uses of 5G Network

On June 6, 2019, the Ministry of Industry and Information Technology (MIIT) granted 5G licenses for commercial use to China Telecom, China Mobile, China Unicom and China Broadcasting Network. China Telecom plans to build a hybrid standalone/non-standalone (SA/NSA) network in 40 cities across the country this year, aiming to take the lead to start SA network upgrading in 2020. China Mobile accelerates 5G network construction and plans to build more than 50,000 5G base stations across the country and provide 5G commercial services to more than 50 cities within 2019. In 2020, it will further expand its network coverage, with a view to providing 5G commercial services to all cities above the prefecture level. China

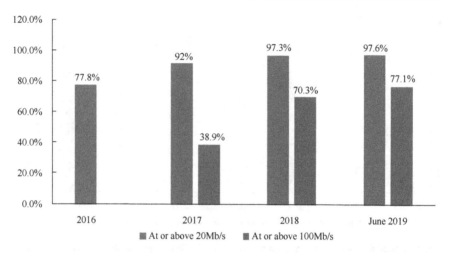

Fig. 1.2 Proportion of broadband users with access rate at or above 20 Mb/s and 100 Mb/s in China from 2016 to June 2019. *Data Source* The Ministry of Industry and Information Technology of the People's Republic of China (MIIT)

Unicom released the "7 + 33 + n" 5G network deployment plan, aiming to achieve continuous coverage in seven cities (urban areas)—Beijing, Shanghai, Guangzhou, Shenzhen, Nanjing, Hangzhou and Xiong'an, and hotspot coverage in 33 cities.

1.2.3.2 The Construction of 5G Industrial Chain Proceeds in an Orderly Way

As 5G technologies and products are maturing, the major links of industrial chain such as systems, chips and terminals have been commercially available. In terms of network construction, China Telecom, China Mobile and China Unicom have built pilot 5G networks previously and are now working on the scale deployment of 5G network for commercial use and the provision of 5G commercial services. In terms of product research and development, significant progress has been made by Chinese enterprises in areas such as medium frequency equipment, 5G base stations and terminal chips. Terminal enterprises have started to launch high-performance mobile phones. Upstream and downstream enterprises actively promote the innovation and development of 5G applications. China Mobile set up its 5G Innovation Center and three Industrial Research Institutes. More than 100 application demonstration projects in nine fields were implemented. China Telecom promoted 5G applications based on the "Three Clouds" architecture—Access Cloud, Control Cloud and Forwarding Cloud in order to meet the demands of low-latency, high-speed data services. It also carried out wide research on industrial Internet. China Unicom carried out effective explorations in fields such as new media, industrial Internet, intelligent tourism, intelligent transportation, intelligent healthcare and intelligent education.

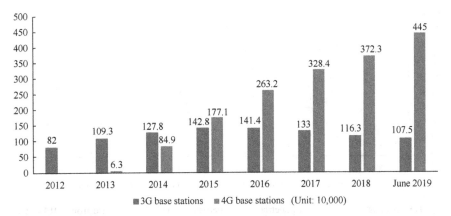

Fig. 1.3 China's construction of 3G/4G base stations from 2012 to June 2019. *Data Source* The Ministry of Industry and Information Technology of the People's Republic of China (MIIT)

1.2.3.3 4G Network Has Achieved Deep Coverage

China has built the world's largest 4G network, with all regions at or above the town level achieving continuous coverage of 4G network and all administrative villages achieving full hotspot coverage. As of the end of June 2019, China had a total of 4.45 million 4G bases stations, accounting for 60.8% of the total number of mobile communications base stations in China. According to *China Broadband Speed Report* released by Broadband Development Alliance, the average download rate on China's 4G networks reached 23.58 Mb/s in the second quarter of 2019. In the early stage of 5G's commercial use, VoLTE remains the major solution of telecom operators' voice services, and China's three major telecom operators accelerate VoLTE deployment. As of the end of 2018, China Mobile's VoLTE users reached 356 million, ranking the first in the world and accounting for 53.4% of the total 4G users. China Telecom has deployed VoLTE across the network. In April 2019, China Unicom carried out commercial trials of 5G in 11 cities of China, including Beijing, Tianjin, Shanghai, Guangzhou and Nanjing, and the nation-wide commercial trials started on June 1. Figure 1.3 shows China's construction of 3G/4G base stations from 2012 to June 2019.

1.2.3.4 The Number of 4G Users in China Ranks the First in the World

With the rapid popularization of "dual-SIM dual-standby" mobile phones, basic telecom operators have launched various preferential packages such as "unlimited traffic package" and "large traffic SIM card" to attract users to buy the second SIM card for their mobile phones, which has stimulated the rapid growth of 4G users. As of the end of June 2019, the number of 4G users in China reached 1.23 billion, ranking the first in the world, with the 4G user penetration rate (proportion of 4G

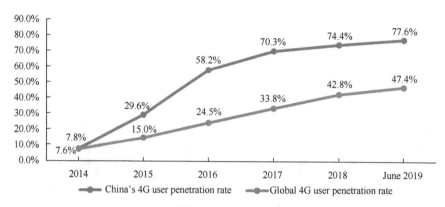

Fig. 1.4 Comparison of 4G user penetration rate between China and the world from 2014 to the second quarter of 2019

users to mobile phone subscribers) of 77.6%, far higher than the global average. Figure 1.4 shows the comparison of 4G user penetration rate between China and the world from 2014 to the second quarter of 2019.

1.2.4 Steady Progress in the Deployment of Space Internet

China has steadily advanced its construction of space information infrastructure. Thanks to technological advancements, policy incentives and further stimulation of market vitality, progress has been made in its low-orbit satellite constellation system, high-orbit high-throughput broadband satellite system and global satellite navigation system (BeiDou Navigation Satellite System).

1.2.4.1 The Low-Orbit Satellite Constellation System Has Entered the Trial Stage

Constellation projects such as "Hongyun" and "Hongyan" were proposed by relevant Chinese enterprises. At the end of 2018, China's first test satellite was successfully launched and substantive progress was made in the technological validation of low-orbit satellite Internet.

The construction project of "Hongyun" system is divided into three stages. According to the schedule, the test satellite would be launched in 2020, and the deployment of entire system would be completed and formal operation would be started during the 14th Five-Year period.

The first communication satellite of "Hongyan" constellation will be used to verify the functions of mobile communications, IoT, navigation augmentation and aviation surveillance while running on orbit. According to the plan, the "Hongyan" constellation project is divided into two phases. In the first phase, a network of 60 satellites will be built to fully cover the "Belt and Road" region. In the second phase, more than 300 satellites will be launched to achieve global coverage.

1.2.4.2 Progress Is Expected to Be Made in the Construction of the High-Orbit High-Throughput Broadband Satellite System

APSTAR-6D is scheduled to be launched before the end of 2019. It will provide high-throughput services in the Asia-Pacific region over the full field of view of the satellite, including the Indian Ocean, the Pacific Ocean, and Australia down to Antarctica. APSTAR-6D is the first high-orbit high-throughput broadband satellite of APT Mobile Satcom Limited (APSATCOM). Three other high-orbit high-throughput broadband satellites will be launched subsequently, which, together with APSTAR-6D, will constitute China's global high-orbit high-throughput broadband satellite system.

1.2.4.3 The BeiDou Navigation Satellite System's Coverage Capacity Continues to Expand

The BeiDou Navigation Satellite System (BDS) is independently constructed and operated by China as an important national space infrastructure that provides all-weather, all-time and high-accuracy positioning, navigation and timing services to global users. As of the end of June 2019, BDS had been iterated to the BDS-3 satellite system, and a total of 46 satellites had been launched into orbit. The BDS-3 system has 21 satellites, which have completed basic system construction and provided services to the world.

1.2.5 Accelerated Expansion of International Communication Facilities

1.2.5.1 International Internet Outlet Bandwidth Has Been Increasing

China's international Internet outlet bandwidth has been developing rapidly. As of the end of 2018, it has reached 8,946,570 Mb/s, more than six times of that of the same period in 2011. But on the whole, China's per capita of international Internet outlet bandwidth is extremely low. Its international Internet outlet bandwidth for each fixed broadband user is only about 0.02 Mb/s. Figure 1.5 shows the growth of China's international Internet outlet bandwidth from 2011 to 2018.

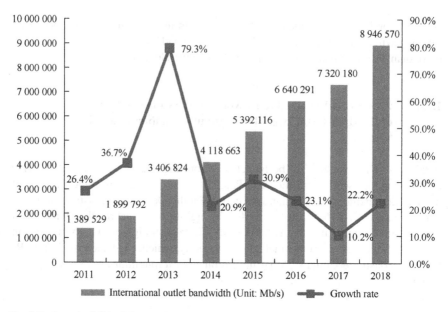

Fig. 1.5 Growth of China's international internet outlet bandwidth from 2011 to 2018. *Data Source* China Internet Network Information Center (CNNIC)

1.2.5.2 Great Progress Is Made in the Construction of International Submarine Optical Cables and Cross-Border Terrestrial Optical Cables

Currently, China has basically built a global transmission network architecture that is composed mainly of multi-directional, high capacity submarine optical cables and cross-border terrestrial optical cables. In terms of submarine cables, China's basic telecom operators have built five submarine cable landing stations and nine international landing submarine optical cables, with the available bandwidth exceeding 100 Tb/s. Meanwhile, they have also actively engaged in the construction of other important international submarine cables. In March 2019, China Mobile started the production of New Cross Pacific (NCP) submarine cable, which is the first directly-connected international submarine optical cable system constructed by China Mobile in China. In terms of terrestrial optical cables, China has 18 terrestrial optical cable border stations and has jointly established cross-border terrestrial optical cable systems with 12 neighbors, with their capacity exceeding 200 Tb/s.

1.2.5.3 There Remains a Gap in the Construction of Overseas Facilities

As of the end of 2018, China's three largest telecom companies had built more than 270 PoPs in more than 35 overseas countries/regions in Asia, Africa, Europe, North America, South America and Oceania. Chinese ICT enterprises have accelerated global deployment of data centers and cloud computing resources. Their data center rooms and businesses span the Asia Pacific region, the United States and EMEA (Europe, the Middle East and Africa), covering all the major Internet markets in the world. There remains a large gap between Chinese CDN enterprises and global leading enterprises in aspects of global service capability and service scale. In Alexa Top 1 K and Top 10 K websites, 70% are American CDN providers, and only 0.5%-3.5% are Chinese CDN providers.

1.2.6   The Pilot Programs of the Universal Telecom Service Boosting Rural Development

1.2.6.1   Sustained Efforts Are Made in the Pilot Programs of the Universal Telecom Service

The universal telecom service is an important means to narrow the digital divide between urban and rural areas/regions and help win the battle against poverty. In April 2019, the fifth batch of the pilot programs of the universal telecom service started to implement. In order to maximize the achievements of the pilot programs of the universal telecom service and make broadband network accessible, affordable and easy-to-use for impoverished users in rural areas, the Ministry of Industry and Information Technology (MIIT) and the State Council Leading Group Office of Poverty Alleviation and Development jointly issued *Notice on Continuously Stepping up Efforts in Network-based Targeted Poverty Alleviation* at the end of 2018. By supporting basic telecom companies to reduce charges for registered poor households and encouraging poor subscribers to increase their use of basic telecom services and various Internet applications, it aims to guide the poor masses to get rid of poverty, become better off, and fully enjoy the dividends of Internet development.

1.2.6.2   The Rural Network Facilities Are Improved Significantly

As of the end of June 2019, the penetration rate of fiber optic network had exceeded 98% in China's administrative villages, and it was 100% in the administrative villages of Beijing, Tianjin, Shanghai, Jiangsu, Zhejiang, Anhui, Shandong, Henan, Guangdong, Chongqing, Yunnan and other provinces (municipalities directly under the Central Government). The number of FTTP ports accounted for 95% of fixed broadband access ports in rural areas, higher than the percentage in urban areas, which was 90%. In order to meet the demands for universal telecom services in rural and remote areas, the Ministry of Industry and Information Technology (MITT) and the Ministry of Finance proposed an "upgraded" scheme for deepening the pilot of universal telecom services. According to the scheme, since the start of 2018, pilot projects would be carried out to promote 4G network coverage in rural and remote areas, and key support would be offered to the construction of 4G network base stations in administrative villages, border areas and island areas. At present, 95% of administrative villages in China have access to 4G network.

1.2.6.3 Network-Based Poverty Alleviation Projects Are Implemented to Lift Impoverished Areas Out of Poverty

The broadband coverage in poor areas of China is greatly improved. As of the end of 2018, more than 97% of total 122,900 listed poor villages across the country were covered by broadband, and the goal of covering more than 90% of poverty-stricken villages with broadband proposed by China's 13th Five-Year Plan had been achieved in advance.

Communication charges are reduced through targeted poverty alleviation. Telecom companies have formulated preferential charge policies and offered "poverty-alleviation packages" for all poor counties and households to greatly reduce the communication charges of the poor masses and encourage them to use broadband services.

Poverty-alleviation applications are promoted to enable the poor regions to share digital dividends. Telecom companies have introduced agricultural technology, agricultural information, educational training, healthcare and rural tourism into poor areas, making broadband network an important channel to open the poor regions to the world and to overcome poverty and achieve prosperity. As Internet services such as distance education, telemedicine and e-government are gradually popularized, high-quality educational, medical and other public service resources in urban areas have been extended to impoverished areas in an accelerated manner.

1.3 Application Facilities Undergoing Rapid Development

1.3.1 The Layout of Internet Data Center (IDC) Maintaining Sound Development

1.3.1.1 The Overall Size of Internet Data Center (IDC) Grows Steadily

China has seen rapid growth in the overall size of IDC, with the compound annual growth rate (CAGR) of more than 30% since 2011. As of the end of 2018, China's total number of in-use Internet data center racks reached 2.042 million, with a year-on-year increase of 23%. Among them, large Internet data centers have witnessed the largest growth. According to incomplete published statistics, China had 41 Internet data center projects under construction and production in 2018. Among them, 26 were projects of constructing large Internet data center with more than 3,000 racks, accounting for more than 60% of the total number.

1.3.1.2 The Demands of Eastern Hot Cities Gradually Shift to Surrounding Areas

After the Ministry of Industry and Information Technology (MIIT) issued *Guiding Opinions on the Construction and Layout of Internet Data Centers* with other four central ministries in 2013, and *Guidelines on the Application and Development of National Internet Data Centers* in 2017, China's layout of Internet data center has been gradually optimized, and newly-built Internet data centers, particularly large and ultra-large Internet data centers, have gradually moved to western regions and surrounding cities of Beijing, Shanghai, Guangzhou and Shenzhen. A large number of new Internet data centers have been built or started operation in Zhangjiakou, Langfang, Ulanqab and Tianjin (surrounding cities of Beijing); Kunshan, Nantong, Suqian and Hangzhou (surrounding cities of Shanghai); Shenzhen-Shantou Cooperation Zone, Dongguan, Zhongshan and Huizhou (surrounding areas or cities of Guangzhou and Shenzhen), among others. The issue of inadequate Internet data centers in China's first-tier cities has gradually been alleviated. For example, there were 24 Internet data center projects constructed and put into operation in 2018 in Beijing, Shanghai, Guangzhou, Shenzhen and their surrounding areas, accounting for 59% of the total number in China. The number of other regional Internet data center projects were 17, accounting for 41% of the total.

1.3.1.3 The Energy-Efficient Management of Internet Data Centers Becomes More Standardized

China has systematically strengthened industrial guidance, review and management of energy-efficient data centers. In February 2019, the Ministry of Industry and Information Technology (MIIT) issued *Guiding Opinions on Strengthening the Construction of Green Internet Data Centers* with other two central ministries, which provided strategic and directional guidance for the construction of green Internet data centers throughout the life cycle from design, procurement, construction, operation, maintenance and transformation. In March 2019, the Ministry of Industry and Information Technology (MIIT) issued *Key Working Plan on Industrial Energy Conservation Supervision in 2019*. According to the plan, special supervision will be conducted on Internet data centers enlisted as key energy-using units by calculating their power usage effectiveness (PUE) according to relevant national standards and inspecting their installation of energy measuring instruments.

In hot cities, the power usage effectiveness (PUE) of Internet data centers is regulated more stringently. Beijing has released *List of New Added Prohibited or Restricted Industries (2018)*, which bars the construction and expansion of Internet data centers across the city (except cloud data centers with the PUE value below 1.4), and prohibits the construction and expansion of data centers in central urban area. In January 2019, Shanghai issued *Guiding Opinions on Strengthening the Overall Construction of Internet Data Centers in Shanghai*, requiring that the PUE value of newly-built data centers should be strictly limited to 1.3 and that of reconstructed

Internet data centers should be strictly limited to 1.4. In April 2019, Shenzhen issued *Notice on Issues Regarding Energy Efficiency Review of Internet Data Centers*, which stipulates that data centers with the PUE value below 1.25 may consume more than 40% of energy, and that those with the PUE value above 1.4 will not receive any support.

1.3.2   Enhanced Capability of Integrating Cloud Computing Platforms

1.3.2.1   Leading Cloud Service Providers Continue to Build Diversified Cloud Computing Platforms

Based on basic cloud computing products, Chinese leading cloud service providers have gradually promoted integration of cloud computing with big data, IoT, AI and other technologies to build a new-generation cloud computing platform facilities with more integrated technologies, more open platforms, better ecology and more diversified business systems. For example, Alibaba Cloud has currently launched more than ten new products that are undergoing public beta testing. Among them, basic products include storage types (e.g., Network Attached Storage, Cloud Photos), databases (e.g., Graph Database) and Internet service products (e.g., IPv6 Translation Service); IoT products include Alibaba Cloud Link WAN, IoT Edge Computing Platform (IECP), Link Visual and IoV Command Center; big data products include data development products and data application products such as DataWorks and Dataphin; AI products include Machine Translation and AI Crowdsourcing. They have enriched Alibaba Cloud's business forms.

1.3.2.2   Foreign Cloud Service Providers Continue to Tap the Domestic Public Cloud Services Market

In May 2019, Microsoft Dynamics 365, Microsoft's third core cloud service, was officially put into commercial use in China. As of then, Microsoft's three key cloud products—Microsoft Azure, Office 365 and Dynamics 365 had all been launched in China. At present, Microsoft Azure and Office 365 have more than 110,000 business customers and more than 1,400 cloud partners in China. According to the data released by International Data Corporation (IDC), Amazon Web Services (AWS), operated by Sinnet, owned 7.2% of the market share in China's public cloud infrastructure market in the first quarter of 2019, ranking the fourth after Alibaba Cloud (43%), Tencent Cloud (12.3%) and China Telecom Cloud (7.3%). In April 2019, AWS Direct Connect landed two new locations in Shanghai and Shenzhen, providing dedicated access to the AWS China (Beijing) region operated by Sinnet. Direct

Connect in Shanghai, operated by NWCD, is live at the GDS Shanghai Third Data Center location[1] and in Shenzhen, also operated by NWCD, is live at the GDS Shenzhen Third Internet Data Center.

1.3.3 CDN Industry Continuing to Grow

1.3.3.1 The Market Size of CDN Industry Has Been Expanding

With the rise of new Internet services such as livestreaming and short videos, the CDN market demands have witnessed continuous growth. China's CDN market size reached ¥ 18.1 billion in 2018 and is expected to reach ¥ 25 billion in 2019, with a growth rate of about 39%. Participants in China's CDN market have increased substantially. As of the end of July 2019, 397 enterprises in China were granted CDN licenses, including traditional CDN service providers, cloud computing service providers, telecom operators, shared CDN service providers and integrated CDN service providers. Among them, 44 have obtained the nation-wide operating license. Cloud computing enterprises are showing an increasing tendency to develop CDN business, and a total of 119 cloud service companies have been granted CDN licenses.

1.3.3.2 CDN and Edge Computing Are Experiencing Integrated Development

With the advent of 5G age, high-traffic applications such as VR/AR, Internet of Vehicles (IoV) and IoT are implemented and more widespread. As the number of Internet connected terminal devices and network traffic are growing exponentially, IoT has caused huge pressure on network access and raised higher requirements for CDN development. Some domestic CDN service providers have begun to launch cloud computing services, and extended their advantages in cloud computing, big data and AI to edge computing to build a cloud-edge-end integrated collaborative computing system. At present, edge computing has been effectively applied into bullet-screen data dissemination in livestreaming. It provides local computing, message receiving and sending, caching and synchronization services for user-owned devices.

1.3.3.3 Domestic CDN Enterprises Are Stepping up the Globalization Pace

With the implementation of the "Belt and Road" Initiative (BRI), Chinese CDN service providers have constantly increased investment in the construction of overseas market. Companies like Alibaba Cloud and Tencent Cloud are widely building data

---

[1]Data Source: https://www.amazonaws.cn/new/2019/aws_direct_connect_shanghai_shenzhen/.

Fig. 1.6 Proportion of different types of domain names registered in China as of June 2019. *Data Source* The 44th "Statistical Report on China Internet Development"

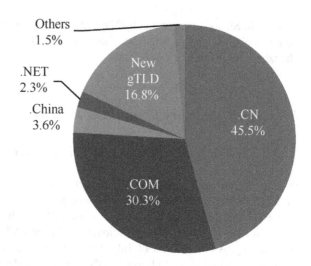

centers and CDN nodes around the world. Currently, Alibaba Cloud owns more than 300 global CDN nodes deployed in 56 availability zones and 12 regions around the world. Tencent Cloud owns more than 200 global CDN nodes deployed in 51 availability zones and 25 regions around the world. Domestic enterprises are shifting their focus to tapping the Southeast Asian market, and many service providers have set up data centers in Southeast Asia.

1.3.4 Increased Holdings of Basic Internet Resources

1.3.4.1 The Market of Domain Name Registrations Grows Steadily on the Whole

As of June 2019, China had a total of 48 million domain names, of which 21.85 million or 45.5% were ended with ".CN", up by 2.9% from the end of 2018, 14.56 million or 30.3% were ended with ".COM", 1.71 million or 3.6% were ended with ".China" (".中国"), and 8.06 million or 16.8% were New gTLDs. Figure 1.6 shows the proportion of different types of domain names registered in China as of June 2019.

1.3.4.2 The Number of Root Server Instances Is Growing

In 2019, CNNIC opened three root server instances in China, namely, F, K and L roots. According to the statistics of CNNIC, China is currently home to 12 root server instances, which are still less than that of the United States. But Chinese institutions

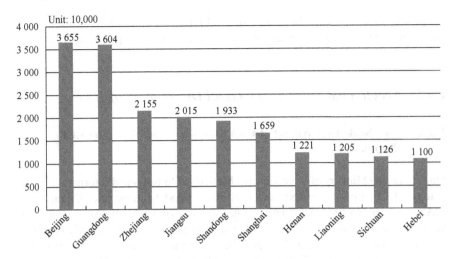

Fig. 1.7 Usage of IPv4 addresses in some provinces (Municipalities Directly Under the Central Government) of China. *Data Source* IPIP.NET

are showing an increasing willingness to introduce root server instances. In 2019, tremendous growth is expected in the number of China's root server instances.

1.3.4.3 The Number of IPv4 Addresses in China Ranks the Second in the World

As of the end of June 2019, China had about 340 million available IPv4 addresses, which were essentially unchanged from 2018, accounting for 9.26% of the total number in the world and ranking the second in the world. Yet, China lags far behind the United States that has a total of 1.606 billion IPv4 addresses. In terms of using zones, Beijing, Guangdong and Zhejiang are the top three regions in the usage of IPv4 addresses. Figure 1.7 shows the usage of IPv4 addresses in some provinces (municipalities directly under the Central Government) of China.

1.3.4.4 The Number of IPv6 Addresses Is Growing

Since the release of *Action Plan for Promoting the Scale Deployment of Internet Protocol Version 6 (IPv6)* at the end of 2017, all regions and authorities in China have earnestly implemented the plan and comprehensively promoted the scale deployment and application of IPv6. The IPv6 notification rate has increased steadily as a result. As of the end of June 2019, the number of IPv6 addresses in China has reached 50,286/32 s,[2] an increase of 14.3% from the end of 2018. This makes China surpass

---

[2]Including Hong Kong, Macao and Taiwan of China.

the United States and rank the first in the world. Besides, China's IPv6 notification rate reached 11.35%, an increase of nearly 20% over the same period in 2018.

1.3.5 Accelerated Scale Commercial Deployment of IPv6

1.3.5.1 The Scale Deployment of IPv6 Enters a Phase of Accelerated Expansion

Since 2018, the Ministry of Education, the Ministry of Industry and Information Technology (MIIT), the State-owned Assets Supervision and Administration Commission (SASAC), People's Bank of China and other ministries have jointly issued *Implementation Opinions on the Action Plan for Promoting the Scale Deployment of Internet Protocol Version 6 (IPv6)*, and 20 provinces (autonomous regions, municipalities directly under the Central Government) including Tianjin, Hebei, Hunan, Liaoning, Jiangsu, Jiangxi, Sichuan, Shaanxi, Yunnan and Zhejiang have issued relevant documents on promoting the scale deployment of IPv6. With these efforts, China's scale deployment of IPv6 has accelerated as a whole.

1.3.5.2 The Network Performance of China's Telecom Operators Remains Stable

As of the end of June 2019, China's three main telecom operators had completed end-to-end IPv6 transformation of mobile broadband access (LTE) networks and started to offer IPv6 services. Their IPv6 services for fixed network users had covered 30 provinces (autonomous regions and municipalities directly under the Central Government) in China. Their total outlet and inlet IPv6 bandwidth had reached 100 GB/s. All the 13 backbone straight points in China had achieved IPv6 interconnection. The total internetwork bandwidth of IPv6 of China Telecom, China Mobile, China Unicom, China Broadcasting Network and China Education and Research Network (CERNET) reached 6.39 Tb/s.

China's main telecom operators have overall stable intra-network and internetwork performance within and between IPv6 networks. In May 2019, the average intra-network delay and packet-loss ratio of IPv6 of China's main telecom operators were 34.45 ms and 0.08% respectively, both approaching the level of IPv4; the average internetwork delay and average packet-loss ratio of IPv6 were 43.63 ms and 0.39% respectively, and their gap with IPv4 was narrowed every month. Figures 1.8 and 1.9 show the intra-network and internetwork performance of China's main telecom operators.

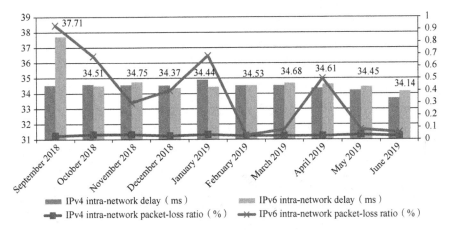

Fig. 1.8 Intra-network performance of China's major telecom operators. *Data Source* The Internet monitoring and broadband speed test platform of the China Academy of Information and Communications Technology (CAICT)

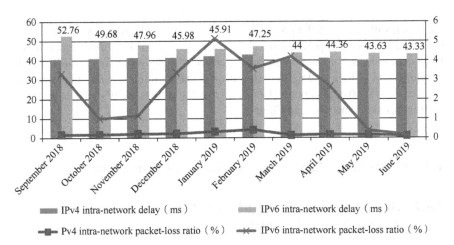

Fig. 1.9 Internetwork performance of China's major telecom operators. *Data Source* The Internet monitoring and broadband speed test platform of the China Academy of Information and Communications Technology (CAICT)

1.3.5.3 IPv6 Traffic Grows Slowly on the Whole

China has witnessed rapidly increasing support over IPv6 among the government and central enterprise websites. In June 2019, Chinese citizens had IPv6 access to 91.2% of provincial-level government portals and 80.2% of central enterprise portals. The websites and applications of key Internet enterprises have accelerated IPv6 upgrading and transformation. Among the top 50 commercial websites and applications in the ranking of the number of domestic users, 80% support IPv6 access. However, there

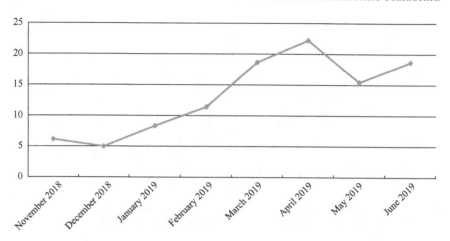

Fig. 1.10 IPv6 traffic (Gb/s) between internet backbone networks in China. *Data Source* The Internet monitoring and broadband speed test platform of the China Academy of Information and Communications Technology (CAICT)

remains insufficiency in the profundity and breadth of IPv6 transformation. To be specific, IPv6 is not supported at many mainstream fixed home network terminals, or there exist risks in software upgrading. These constraints seriously jeopardize IPv6 development in China, and result in slow growth in IPv6 traffic and the small size and proportion of IPv6 application in China. Figure 1.10 shows the IPv6 traffic between Internet backbone networks in China.

1.4   Accelerated Pace of New Facilities Construction

1.4.1   Accelerated Facility Deployment Driven by Rising Demands in IoT Industry

1.4.1.1   Remarkable Results Achieved in NB-IoT Network Construction

China has built the world's largest NB-IoT network, and China Telecom, China Mobile and China Unicom have built more than one million NB-IoT base stations for commercial use. With the high-quality spectrum resource of 800 MHz, China Telecom expanded its base stations to 400,000 in 2018, achieving further deep coverage in China. China Mobile built around 300,000 NB-IoT base stations in 2018, achieving continuous coverage and commercial use of NB-IoT in nearly 350 cities. China Unicom deployed 300,000 NB-IoT bases stations for commercial use in 2018.

1.4.1.2 The Popularization of IoT Applications Is Accelerated

China's IoT industry has maintained rapid growth. According to the statistics of the Ministry of Industry and Information Technology (MIIT), as of the end of 2018, China's three main telecom operators had 671 million cellular IoT users, with a net increase of 400 million over the previous year, accounting for more than half of the total number in the world. In 2018, China's IoT industry reached ¥ 1.2 trillion, 80% of the target to be achieved at the end of the 13th Five-Year Plan. The industry in Jiangsu, Zhejiang, Guangdong and several other provinces has all exceeded ¥ 100 billion. At present, IoT applications are penetrating into the whole industry process ranging from R&D, manufacturing, management and services. In industries such as agriculture, transportation and retail, pilot programs of IoT application are also carried out in an accelerated manner. The market potential of consumer-oriented IoT applications will be gradually released. Smart home, wearable devices for health management, intelligent door locks, intelligent vehicle terminal and other consumption fields have witnessed continuous rapid growth in IoT applications. As smart cities have entered a stage of comprehensive construction, large-scale application of IoT will be a future trend.

1.4.2 Phased Progress in the Construction of Network Facilities of Industrial Internet

1.4.2.1 Steady Progress Is Made in the Construction of Industrial Internet Network

2018 was China's first year for the comprehensive construction of industrial Internet.

(1) The top-level design was gradually improved. The Alliance of Industrial Internet (AII) released *Industrial Internet Networking & Connecting White Paper* and *Standardization System for the Industrial Internet (Version 2.0)*, which clearly defined the architecture of the industrial Internet network in technology and standard.

(2) Network upgrading and transformation produced initial results. Basic telecom operators started to build high-quality backbone networks beyond enterprises. NB-IoT, which enabled WAN connection of low-power device, became available nationwide. Manufacturing enterprises actively utilized the passive optical network (PON), edge computing, IPv6 and other new technologies to reconstruct the enterprise-wide network.

(3) The exploration of new network technology accelerated. The Alliance of Industrial Internet (AII) built more than ten network test beds for Time-Sensitive

Networking (TSN), edge computing, 5G and other new network technologies. All parties in the industry jointly built many network innovation laboratories and launched more network solutions that catered for the demands of the transformation and upgrading of manufacturing enterprises.

1.4.2.2 Preliminary Results Are Produced in the Construction, Application and Promotion of the Identification and Resolution System

(1) Phased progress has been made in the construction of the identification and resolution system. Five national top-level nodes have been released and put into trial, and the top-level layout of "East, West, South, North and Center" has initially taken shape. In addition, 11 secondary nodes have started operation, covering fields such as high-end equipment, engineering machinery and aerospace.
(2) China has built strong independent innovation capabilities. The integrated identification technology scheme has been successfully validated. A framework of the standard identification and resolution system has been built. Relevant software of the identification and resolution system has been developed and is integrated with new-generation information and communication technologies such as AI and blockchain in an accelerated manner.
(3) Identification applications are taking on a path of innovative development. The identification and resolution technology has been applied into the life-cycle management of industrial products, equipment asset management, supply chain management, product traceability and other fields. Through coordinated efforts of the "Government-industry-university-research-employer" Cooperation, an ecosystem of the identification and resolution industry has taken initial shape.

1.4.2.3 Industrial Internet Platforms Are in the Initial Stage of Development

China's industrial Internet platforms are still in the initial stage of development and are deployed in some vertical market segments.

(1) Large enterprises have accelerated platform-based transformation and constantly launched platform products. Advanced manufacturing enterprises have transformed their digital transformation experience into platform services. Equipment and automation enterprises have innovated their service models by relying on their industrial equipment and experience. IT enterprises have exerted their advantages to extend their platforms to the manufacturing field. Internet enterprises have furthered developed industrial solutions based on their cloud services.
(2) Chinese enterprises are developing technological products and solutions for data acquisition, integration and analysis. A great many practices are done through industrial Internet platforms, including equipment data acquisition,

edge computing analysis, development and invocation of industrial mechanism models and microservices, industrial big data storage analysis, and development and deployment of industrial apps. Besides, they are also exploring innovative enabling practices such as equipment health management and manufacturing capacity trading.

(3) Software cloudization and industrial Apps have developed rapidly. Software enterprises have strengthened their capacity of software cloudization and data aggregation and processing, and developed more than 300 industrial apps for different use cases.

1.4.2.4 Positive Progress Is Made in the Construction of the Industrial Internet Security System

China's industry has made positive progress in establishing security frameworks and standard systems and developing technological products. In terms of security frameworks, the Alliance of Industrial Internet (AII) has released a series of framework guidelines and solutions, including *Security Framework of Industrial Internet* and *Typical Solutions for Industrial Internet Security*. In terms of standard systems, the framework for industrial Internet safety standard system has been established. The general requirements for industrial Internet security and the standards of platform security and data security have been further improved. In terms of technological capability, the technological support systems at the national, provincial and enterprise levels have been initially established, and collaborative efforts are made by key laboratories, professional institutions and security enterprises in technological research and development and integrated application. In terms of industrial ecology, as China's cybersecurity demand is rising, more safety products and solutions are generated and security companies are growing rapidly.

New infrastructure represented by 5G, AI, IoT and industrial Internet has become the pillar of high-quality economic development. China should promote the scale deployment of 5G network, improve its IoT access capabilities, and promote the intelligent transformation, upgrading and evolution of networks. In particular, it should strengthen network construction in the central and western regions and rural areas, bring local information infrastructure construction onto the fast track in an all-round way, and build a new infrastructure system that suits the needs for integrating digital and real economies. With these efforts, the upgrading and application of information infrastructure will become new growth points for the development in all regions and lay a solid foundation for building China's strength in cyberspace.

Chapter 2
Development of Network Information Technology

2.1 Outline

Core technologies are the cornerstone of informatization and a pillar of the nation. General Secretary Xi Jinping stressed, "To build China's strength in cyberspace, we must have our own technology, excellent technology." In 2019, China has further deepened and expanded innovation and industry-university-research collaboration in network information technology and research on key technologies. As a result, breakthroughs are made in basic technologies. Cutting-edge technologies present distinct features of cross-border integration, systematic innovation and intelligence guidance. The construction of new technologies, new processes and new platforms is accelerating. A number of independent innovation achievements have reached the world leading level.

New progress has been made in the research and development of basic and universal technologies. The focus of industrial development has shifted from application-oriented innovation to technological innovation. Efforts have been constantly stepped up in the research and development of basic technologies. Progress has been made in the new-generation exascale supercomputer, IoT OS, 5G base-band chips and other sub-fields. However, it is also noticeable that the technological innovation in fields such as basic chips, operating systems and industrial software is characterized with high systematicness, accumulativeness and progressiveness. Persistent and concentrated efforts are thus required to comprehensively promote technological innovation in these fields.

Breakthroughs have been made in cutting-edge technologies. The 7 nm process technology has been deployed in AI chips, and the quantum key distribution protocol has reached the international leading level. Especially with the commercial deployment of 5G network, edge computing technologies have been gradually applied and an ecosystem of edge computing has been established. Besides, the integration of virtual reality and traditional industries has accelerated.

© Publishing House of Electronics Industry 2021
Chinese Academy of Cyberspace Studies, *China Internet Development Report 2019*, https://doi.org/10.1007/978-981-33-6930-6_2

Increasing efforts have been put in the research and development of core technologies in cyberspace affairs. Since 2019, the CPC Central Committee has incorporated the task of seeking breakthroughs in key and core technologies into important policy documents, including *Guidelines of the CPC Central Committee and the State Council on Supporting Shenzhen in Building a Pilot Demonstration Area of Socialism with Chinese Characteristics* and *Framework Plan for the New Lingang Area of China (Shanghai) Pilot Free Trade Zone*, for the purpose of comprehensive planning and promotion of relevant work. Research institutes such as the Chinese Academy of Sciences, Tsinghua University, Zhejiang University and the University of Science and Technology of China (USTC) have carried out continuous research in basic theories, basic technologies, advanced technologies and other areas by giving play to their own research advantages. Alibaba, Tencent, Baidu, Huawei and other companies have further engaged in the innovation of basic technologies and cutting-edge technologies, and produced a number of achievements. Research and development organizations are playing an increasingly important role.

2.2 Steady Progress in Basic Network Information Technology

Over the past year, steady progress and constant breakthrough have been achieved in the basic network information technology. China's high-performance computing capability has been leading the world. Breakthroughs have been made in some integrated circuits and software technologies. Progress has been made in resolving the issue of "lack of domestic circuits" in China.

2.2.1 Strengthened Capability of High-Performance Computing

At present and for a long time to come, the breakthroughs and applications of network information technology will be of great significance in the technological strategies of all countries. High-performance computing technologies, by virtue of their strong numerical computation and data processing capacity, play an important role in helping achieve breakthrough and wide application of the new-generation network information technology. High-performance computing technologies represented by supercomputer have been widely integrated into every aspect of scientific and technological innovation, economic development and social life, and been comprehensively applied into areas such as astrophysics, climate science, oceanography, aerospace, life science, AI and big data. They are playing an increasingly important role.

2.2.1.1 China Maintains Its Competitive Advantage in the Number of Supercomputer

As an important carrier of high-performance computing, supercomputer reflects a country's position as a strong power in the global competition in network information technology. The research and development of supercomputer started early in China. After decades of development, it has achieved many independent innovative breakthroughs and developed a series of supercomputer systems such as "Galaxy", "Tianhe" and "Shuguang". China has thus jumped out to the lead and formed its advantages in this field. In June 2019, TOP 500, an international organization for supercomputer performance evaluation, released the list of top 500 supercomputers in the world. Among them, there were 219 in China, 116 in the United States, 29 in Japan, 19 in France, 18 in the UK and 14 in Germany.[1] China has sustained its competitive advantage in the number of supercomputer. The Chinese supercomputer "Sunway TaihuLight" ranks the first in China and the third in the world, with a LINPACK benchmark rating of 93 petaflops. All its core components, including the processor, have achieved localized production. Meanwhile, it is important to note that there remain gaps in the performance indicators between Chinese supercomputer and the "Summit" supercomputer in the United States, which ranks the first in the world.

On the whole, China has steadily improved its high-performance computing technologies and actively developed E-class computing capabilities. "Shenwei", "Tianhe-3" and "Shuguang", three prototypes of the new-generation E-class supercomputers with independent intellectual property rights, have completed research and development and are gradually applied into many fields, indicating that China's E-class supercomputer will soon enter the stage of substantive research and development.

2.2.1.2 Intelligent Computing Becomes an Important Use Case of High-Performance Computing

As deep neural networks mature and big data technologies thrive, supercomputing and intelligent computing are integrated historically, and high-performance computers are applied from scientific computing into big data and machine learning. In August 2019, the Brain-inspired Computing Research Center of Tsinghua University developed the world's first heterogeneous fusion brain-like chip, "Tianjic". The chip consists of 156 Fcores, containing approximately 40,000 neurons and 10 million synapses. Computer-science-oriented and neuroscience-oriented approaches are integrated on one platform, which effectively facilitates the research and application of artificial general intelligence. Based on the research results, the paper *Towards Artificial General Intelligence with Hybrid Tianjic Chip Architecture*, was published

---

[1] http://www.top500.org/lists/2019/06/.

in *Nature*, an authoritative science and technology magazine in the UK, which marked a breakthrough in the publication of a paper in *Nature* by China in the fields of chips and artificial intelligence.

2.2.2 Progress Made in Software Technology

Software is both the soul of the new-generation information technology and the key pillar for building China's strength in cyberspace and a manufacturer of quality. China is an important growth pole for the development of the global software industry. In 2018, the operating income of China's software industry reached ¥ 6.3 trillion, up 14.2% year-on-year, and the total profit reached ¥ 807.9 billion, up 9.7% year-on-year. Particularly, the innovation capacity of software industry has been greatly improved. In 2018, the R&D intensity of software industry reached 10.4%, and the copyright registrations for computer software exceeded 1.1 million, with innovation achievements constantly produced.[2] From the development trend of the software technology system, operating systems and industrial software have become the key support for the development of the network information industry and advanced manufacturing industry and the key areas of current concerted efforts. Software-defined anything, also known as SDX, represents a new trend of the coordinated development of software and hardware in the field of information technology, which is highly valued by all relevant parties.

2.2.2.1 Domestic Operating Systems Accelerate Development

Operating systems are indispensable basic software. For a long time, the Chinese operating system market is dominated by foreign world. As of June 2019, the penetration rate of Android on mobile devices reached 78.23%, and that of Windows on desktop reached 82.55%.[3] All relevant parties have stepped up efforts to independently develop operating systems and their applications in some fields have reached a certain scale. In the field of emerging IoT technologies, more than one solution has been produced and many projects have been implemented.

(1) Positive progress is made in the development of desktop and mobile-end operating systems

At present, China's desktop operating systems are mainly developed based on Linux, an open-source operating system. Their overall performance is continuously improved, and a great many operating systems are developed, including "NeoKylin", "Kylin" and "DeepIn". In July 2019, a new version of "DeepIn" was released, which

---

[2]ThePaper.cn: https://www.thepaper.cn/newsDetail_forward_3789879, last access time: September 4, 2019.

[3]Data Source: https://gs.statcounter.com/, July 2019.

rose to the 11th place in the list of Linux distributions. For mobile-end operating systems, many research institutes and enterprises have engaged in their development and research, producing a large number of achievements. In April 2019, the Chinese Academy of Sciences released FactOS v1.0, an intelligent operating system that supports multiple processors and accelerators, programming frameworks, data sets and neural network models, and can be widely applied into deep learning and other use cases. Huawei launched the Ark compiler in the same month, which was officially open source in August. The compiler can be used for static compilation of Java source code for Android applications, providing a domestic autonomous and controllable alternative for some core components of Android operating systems. In August 2019, Huawei officially released its self-developed "HarmonyOS", a new microkernel-based, distributed operating system. It features low latency and provides on-demand scalability. Huawei's independent research on its "kernel and application framework" is expected to be completed in 2020.

(2) China's cloud operating systems are competitive around the globe

More than 50,000 and 6,000 sets of "Kylin", a domestic operating system based on Linux, have been implemented in the power industry and aerospace industry respectively.[4] Dragonfly, an image distribution system released by Alibaba, was officially included in the world's top open-source community Cloud Native Computing Foundation (CNCF) as one of its Sandbox Level Projects, and its cloud native value has been recognized in the industry. Huawei has released FusionSphere, a cloud operating system based on OpenStack architecture. The whole system is designed and optimized for cloud and provides powerful conversationalist functionality and resource pooling management, as well as rich cloud-based service components and tools. Alibaba Cloud launched "Apsara 2.0", a new-generation cloud computing operating system for IoT, at Alibaba Group's Apsara Conference held in Hangzhou in 2018. The system meets the computing needs of tens of billions of smart devices and can be applied into IoT and supercomputing case, thus promoting the intelligentization from means of production to means of livelihood. Besides, PingCap has independently developed TiDB, an open-source, distributed relational database as reliable and secure as commercial databases.

(3) IoT operating systems develop rapidly

Driven by the commercial uses of 5G technology, various parties are pushing forward the development of IoT operating systems. Huawei has launched a series of projects for the commercialization of intelligent parking, intelligent water meters, intelligent lighting and other technologies based on its IoT operating system LiteOS. Alibaba has unified the development framework and standards of its IoV operating system "AliOS", and released a lightweight IoT operating system "AliOS Things". The

---

[4]https://news.changsha.cn/cslb/html/111874/20190619/46527.shtml.

Chinese Academy of Sciences has released RVOS, an operating system designed for RISC-V, an open-source Instruction Set Architecture (ISA) based on Linux and FreeRTOS. Steady progress has also been made in other platforms such as RT-Thread and SylixOS.

(4) Cloud desktop technology promotes the smooth migration of ecosystem

In order to improve the ecology of domestic operating systems and provide better user experience, desktop manufacturers such as Huawei, Sangfor Technologies and Centerm WeiXun have launched their own hybrid architecture solutions. These solutions have promoted the smooth operation of x86 cloud desktops and applications on domestic operating systems by using desktop virtualization and virtualization technologies. In terms of key technologies, both the server CPUs and desktop transmission protocols adopt the heterogeneous architecture. Different from x86 desktops that usually adopt the Virtual Desktop Infrastructure (VDI) model, domestic desktops use the Intelligent Desktop Virtualization (IDV) model. For the latter, the computing tasks of servers are assigned to terminals in order to ensure the smooth operation of desktops.

2.2.2.2 Industrial Application of Software Definition Is Deepening and Expanding

The core value of Internet is connection, which further relies on software technology. With the further development of AI, IoT, big data and other technologies, Software Defined Network (SDN), Software Defined Storage (SDS) and Software Defined Computing Pattern (SDCP) have continued to develop, the "human-machine-things" interaction has accelerated, and the concept "Software Defined Anything" (SDX) has gained increasing attention. The heart of soft definition is to give better play to the dominating role of software in controlling hardware resources and providing more open, flexible and intelligent management and control services through hardware virtualization and programmability of management functions. In April 2019, the Chinese Institute of Electronics (CIE) established the Software Definition Promotion Committee to strengthen industry-university-research collaboration in software definition. From the development practices, software definition has increasingly sped up expansion and made remarkable achievements in many fields. For example, as of June 2019, "Tianzhi-1", China's first software-defined satellite developed by the Chinese Academy of Sciences, has completed more than 10 experiments in orbit, including satellite-rocket separation imaging, measurement and control upon autonomous request and space target imaging, by means of software annotation. Huawei Fusion Storage, the first distributed storage system in the industry to deliver data center-level convergence, can break down data barriers between different data types, life cycles, spatial distribution and business types to achieve the goal of "One Internet Data Center One Storage".

2.2.2.3 Progress Is Made in the Development of Industrial Software, Particularly Computer-Aided Software

Industrial software is an important basis and core support of intelligent manufacturing. On the whole, despite huge gap between industrial software enterprises in China and those in other countries, the new-generation information technology such as cloud computing, IoT and big data has been further integrated with the industry, which has created new opportunities for China's industrial software development. Over the past year, great progress has been made in China's computer-aided software. For example, Zwcad Software integrated computer-aided design (CAD), computer-aided engineering (CAE) and computer-aided manufacturing (CAM) tools to develop autonomous and controllable domestic industrial software. GstarCAD Software directly embedded the collaborative AutoCAD design system into CAD design software, which improved the efficiency of multi-specialty coordination through remote collaboration and achieved transition from single-point efficiency improvement to overall efficiency improvement. In June 2019, LiToSim-Software released LiToSim, a domestic independent industrial simulation software applied into the construction of China's high-speed rail. The enterprise-level R&D management platform CODING released Cloud Studio, a cloud-based online collaborative programming platform, to integrate services such as project management, code development and cloud hosting. It performs better in terms of software development, rapid computing and cloud collaboration, making "cloud-based working" possible. However, China is lack of technological strength in Electronic Design Automation (EDA), and domestic EDA solutions are insufficient to meet the demands of informatization development.

2.2.3 Continuous Development of Integrated Circuit Technology

Integrated circuits, known as the "food" for IT industry, are playing a strategic, fundamental and leading role in the industry. Speeding up the development of integrated circuits is a strategic choice for seizing the opportunity of the new round of scientific and technological revolution and fostering new drivers of economic growth. It is also an essential way to deepen the supply-side structural reform and promote high-quality economic development. On the whole, China is witnessing active innovation in network information technology, booming digital economy and accelerated development of IoT and intelligent manufacturing. The continuously rising demands for integrated circuits have greatly spurred technological breakthroughs and industrial development of integrated circuits.

2.2.3.1 Single-Point Breakthroughs Are Made in the Research and Development of Basic Chips

(1) Computing chips of various architectures develop in parallel

There are currently many different architectures of central processing unit (CPU) chips in China, including x86, MIPS, ARM, Power and Alpha. Through proactive efforts of Haiguang, Loongson, Huawei Hisilicon, Sunway and other enterprises, relevant products have been substantially applied into the server market and mobile-end market. Some enterprises have started to develop GPU and other chips mainly by building SoC systems with foreign IPs. They include Loongson's 2H chip, MPRC's Tiandao chip, Huawei Hisilicon's MALI, which are however far behind the international advanced level.

(2) Memory chip enterprises vigorously develop autonomous key technologies

A great many leading enterprises, including YMTC, Innotron Memory and GigaDevice, have actively made breakthroughs in memory technologies. In terms of Dynamic Random Access Memory (DRAM), Innotron Memory's 19 nm 8 Gb LPDDR4 memory chips are expected to be put into volume production in the end of 2019. In terms of NAND flash memory chips, in September 2019, YMTC started volume production of its 256 GB 64-layer 3D TLC NAND flash memory chip, and released Xtacking 2.0, a 3D flash memory technique, raising the speed of NAND flash memory up to 3.0 GB/s. In general, China's memory chips have maintained a sound growth momentum, but continuous innovation and breakthrough are still required to build a large-scale industry.

(3) The design of communication chips, particularly 5G baseband chips of mobile intelligent terminals, reaches the world advanced level

Under the ARM architecture license, Hisilicon and Spreadtrum have developed their own core chips, produced 64-bit multi-core SOC chips with multi-mode multi-frequency LTE, and achieved the 7 nm process technology. Large gap still exists in the development of RF chips between China and the international level. Leading enterprises in this field include Unisoc, Vanchip, Smarter Micro and Huntersun Electronics. There are only four companies in the world making baseband 5G chips. Except those developed by Qualcomm, all the baseband 5G chips are independently designed by China. They include Huawei Balong 5000, MediaTek Helio M70, and Chunto 510 released by Unisoc in 2019. With the characteristics of high integration, high performance and low power consumption, Chunto 510 can be integrated with many 5G core technologies.

2.2.3.2 Progress Is Made in Advanced Technologies

The integrated circuit manufacturing processes have been continuously improved. For example, the 12-inch production line uses the 65–14 nm manufacturing process, the 8-inch production line uses the 0.25 μm–90 nm manufacturing process, and the 6-inch production line uses the 1.0–0.35 μm manufacturing process, all of which have achieved scale production. For key equipment, breakthroughs have been made in self-developed key equipment of integrated circuits, with their overall level reaching 28 nm. Besides, 16 types of key equipment, including ion implanters, etchers, physical vapor deposition (PVD) and chemical mechanical polishing (CMP), have been verified on large production lines and put into operation. Among them, Chinese etchers are now able to compete with similar products in the international market and are extensively applied into advanced production lines. For example, AMEC's 5 nm etchers have entered the supply system of international mainstream enterprises. The testing equipment in China mainly consists of the analog/digital analog hybrid testing equipment, nearly 80% of which are produced in China. Great progress has been made in deposition equipment, with China's self-developed PVD equipment having entered the evaluation stage of 14 nm process.

2.2.3.3 China's Open Source Hardware Follows Closely Behind the International Advanced Level

With great scalability and convenience, open source hardware has advantage in technological innovation. Major products of domestic open source hardware are RISC-V and MIPS. Cutting-edge research on RISC-V, an open source instruction set architecture (ISA), is advanced steadily, which develops in parallel with foreign open source communities. RISC-V-based products are developed by relevant enterprises, one of which is XuanTie910, a RISC-V processor released by Alibaba's chip subsidiary Pingtouge in August 2019. It can greatly reduce the design and manufacturing costs of high-performance end-to-end chips, and is one of the highest-performance RISC-V chips. MIPS R6 is put on open source, but is not compatible with the early version of MIPS. Its ecological construction is underway.

2.2.3.4 Domestic Chips Boost the Performance of Servers

Taishan server, running on the Kunpeng 920 processor, has been extensively launched in the domestic market. With many computing cores integrated, its overall performance has reached that of the current mainstream x86 servers. Tianyue SR117220 server, running on the Loongson 3B3000 processor, is proved to get close to intelE5, a mainstream processor in the market, in overall performance. With the efforts of Chinese chip manufacturers and design optimization by whole machine manufacturers, rapid progress is expected to be made in the performance of the servers running on domestic CPUs such as Sunway and Hygon.

2.3   Active Innovation in Cutting-Edge Technologies

The world has entered an unprecedented period of extensive and active scientific and technological innovation. Cutting-edge technologies represented by AI, quantum information, IoT, blockchain and 5G are emerging endlessly and applied in an accelerated manner. Through strengthening policy planning and goal guidance, cutting-edge technologies such as 5G and cloud computing have produced remarkable achievements and accelerated practical applications. At the same time, China has made constant efforts to catch up with global frontier technologies and formed its superiority in sub-fields such as AI and quantum information.

2.3.1   Continuously Deepened AI Development

AI is a strategic technology heralding the new round of scientific and industrial revolution and industrial change which has a "lead goose" effect with a strong stimulating nature. Under the impetus of mobile Internet, big data, supercomputing, sensor networks, neuroscience, and other new theories and new technologies, AI has developed rapidly and presented the new characteristics of deep learning, interdisciplinary convergence, human-machine cooperation, group intelligence development, autonomous operation, etc. It is currently exerting a major and profound influence on economic development, social progress, and the structure of international politics and economy. Accelerating the development of the new- generation AI is an important strategic handhold for China to gain the initiative in global scientific and technological competition. It is an important strategic resource that drives the country's leapfrog development in science and technology, industrial optimization and upgrading, and a comprehensive leap ahead in productivity.

2.3.1.1   The Development of AI Chips Is Accelerating

As the basis of algorithm running, chip receives increasing attention. AI chips and relevant hardware products have been successfully developed. In October 2018, Huawei officially released two universal AI chips: the 7 nm-based "Shengteng 910" and the 12 nm-based "Shengteng 310". "Shengteng 910" is currently the world's largest single-chip computing density AI chip, with a semi-precision power of 256 TFLOPS. "Shengteng 310" features a low-power AI case with 8 TFLOPS half-precision computing power and a maximum power consumption of 8 W. Besides, many AI enterprises have also launched AI chips for particular use cases. For example, in January 2019, Unisound announced its development of three AI chips: "Swift Lite", a second-generation IoT voice AI chip; "Dolphin", a multi-mode AI chip providing image and voice computing for smart cities; and "Leopard", a vehicle-level multi-mode AI chip for intelligent traveling. Besides, in May 2019, YITU launched Questcore, a cloud-based AI chip for deep learning and visual reasoning.

2.3.1.2 Progress Is Made in a Variety of AI Algorithms

As an important fundamental task in Natural Language Processing (NLP), semantic role tagging (SRL) has gradually become a focus of research. An increasing number of researchers have begun to construct end-to-end SRL models without syntactic input. In March 2019, Cloudwalk and Shanghai Jiaotong University proposed a unified end-to-end SRL method based on an original DCMN algorithm. This method produces higher question-answering accuracy than middle school students in a large deep reading comprehension task, and becomes the world's first NLP model that scores higher than humans in reading comprehension.

Semantic segmentation, as one of the most important tasks in computer vision, has been studied extensively in the past years. Modern approaches for semantic segmentation usually employ dilated convolutions in the backbone to extract high-resolution feature maps, which causes heavy computation complexity and memory footprint. In view of this, in April 2019, the Chinese Academy of Sciences and Deepwise AI Lab proposed to replace the dilated convolutions with a joint upsampling module named JPU (Joint Pyramid Upsampling), which effectively reduced the computational complexity and memory footprint with improved performance.

Besides, innovation has been made in underlying neuron models. Researchers from Tsinghua University, Google and ByteDance propose the Neural Logic Machine (NLM), a neural-symbolic architecture that uses a forward chaining model to effectively reduce logic complexity and applies neural network into logic reasoning.

2.3.1.3 AI Open Source Software Is Thriving

As many AI enterprises have established open-source machine learning frameworks, an ecosystem is emerging in this field and the threshold of AI application in different areas has been lowered. From December 2018 to May 2019, Alibaba launched a series of open source software, including X-Deep Learning (XDL)—an algorithm framework for advertising business, PAI V3.0—a platform for machine learning, and Mobile Neural Network (MNN)—a lightweight DNN (Deep Neural Networks) inference engine. In May 2019, SenseTime released a series of innovative AI open source software platforms, including the educational experiment platform SenseStudy AI and the comprehensive solution Sensear Avatar, to constantly improve the impact of AI technology on different industries.

Deep reinforcement learning has been one of the most concerned topics of research in the AI field in recent years. It has produced fruitful achievements in areas such as games and robot control. In January 2019, Baidu officially released PaddlePaddle deep reinforcement learning framework PARL and opened sourced PARL's complete training code that won the NeurIPS 2018 Reinforcement Learning Event. PARL is more scalable, reproducible and reusable than the existing enhanced learning tools and platforms, which supports massively parallel and sparse features and enables rapid verification of industrial-grade use cases.

2.3.1.4 AI Is Successfully Applied into Many Fields

As the application of AI in automatic driving has continued deepening, well-known automobile manufacturers and Internet enterprises have invested in its research and development. In July 2019, China's first L4 passenger car front-loading production line built by Baidu and FAW Hongqi was officially put into production, and the first batch of L4-class self-driving passenger cars took the lead to land in Changsha. Baidu also announced the release of Apollo 5.0, which upgraded both Apollo Open Platform and Apollo Enterprise. The Apollo Open Platform upgraded 17 core capabilities and made its data pipeline accessible to developers, empowering them with six data capabilities, including intelligent data collection, open synthetic dataset, large-scale cloud training, custom simulation validator, open data application set and seamless compatibility with the Apollo Open Source Platform. A whole process integrating data collection, training, verification and full release was established.

Computer vision technology gives rise to many applications that grow quickly. Deep Awakening Technology has cut into the AI battlefield with face recognition technology and provided solutions for security monitoring, finance, real estate, schools, hospitals and other fields. Shanghai Pudong Development Bank deployed YITU's facial recognition platform for video teller machines (VTM) and mobile banks. DeepGlint has released a series of intelligent hardware products based on computer vision technology, including DeepGlint Smart Eye Camera, DeepGlint FoveaCam, Haomu Behavior Analyzer, and DeepGlint Telepresence Robot. It has also provided various application platforms such as the full target structured system, the video and image resolution system, the face recognition system and the mobile control system.

2.3.2 Steady Progress in Edge Computing

As a new data computing architecture and organizational form, edge computing has expanded the scope of data computing from cloud centers to device ends, making it more convenient to provide intelligent services. With the advent of 5G age, the amount of data generated by network edge devices has grown rapidly, which calls for higher data transmission bandwidth and faster data processing. As a result, edge computing has developed rapidly and accelerated its entry into the market. In 2019, China's edge computing developed steadily thanks to the evolution of key technologies such as operating systems, algorithm platforms as well as security and privacy.

2.3.2.1 The Design of Edge Computing Operating Systems Is Evolving into Specialized Use Case

Edge computing operating system needs to manage different types of computing resources downwards and deal with massive heterogeneous data and loads of various applications upwards. To ensure the reliability of the computing tasks and maximize the use of resources has become the research and development direction of edge computing operating system. In May 2019, the NIIC released NECRO, its self-developed real-time operating system for industrial Internet, to meet the data integration and deep modeling demands in the industrial field and lower the threshold of using edge computing. In August 2019, Zhongke Haiwei initially completed its research and development work on "Seaway", an edge computing operating system with independent intellectual property rights in China. Like an ultra- lightweight object-oriented Android OS, it allows the development, loading, execution and unloading of third-party applications. Besides, edge computing operating systems for smart home, intelligent connected vehicles and intelligent robots are developed currently.

2.3.2.2 Edge Computing Platforms Are Developing Towards Cloud-Edge-End Integration

With the advent of 5G age and the development of IoT, the future computing tasks will not only be carried out in large data centers, but also on the "cloud-edge-end" integrated continuous spectrum. Since 2018, Domestic enterprises have actively deployed cloud-edge-end integration. Alibaba Cloud released Beta Link Edge, an edge computing product, and worked hard to build a cloud-edge-end integration platform. Baidu Cloud opened source its intelligent edge computing platform OpenEdge, which extended the application of cloud computing to user's case and provided temporarily offline and low-latency computing services. In January 2019, Baidu launched Baidu Intelligent Edge (BIE), an intelligent edge computing product enabling the integration of cloud and edge. With one-click release and sensor-less deployment functions, the speed of intelligent iteration has been improved. In July 2019, KubeEdge, an intelligent edge project open sourced by Huawei Cloud, received the Leading Open Source Technology Innovation Award, becoming the first official project of the Cloud Native Computing Foundation (CNCF) in the intelligent edge field. In July 2019, Beijing Youfan Technology released its IoT service platform QingCIIoT and the edge computing platform EdgeWize to promote the cloud-network-edge-end integration. In August 2019, UNISOC released Tiger T710, a high-performance AI-enabled edge computing platform, to provide high-performance and low-power technology base for various applications. In the latest AI Benchmark chip testing list released by the Swiss Federal Institute of Technology Zurich, the UNISOC Tiger T710 topped the leaderboard with an outstanding score.

2.3.2.3   The Privacy and Security Design of Edge Computing Is Entering the Initial Stage

Edge computing makes computing closer to users, eliminates the need to upload data onto the cloud and reduces the risk of privacy leakage. However, as edge computing devices are usually located near the user side, they are more likely to be attacked on the transmission path. In recent years, some emerging security technologies, such as hardware-assisted trusted execution environments (TEEs), have been applied into edge computing, but the research results are still scarce. China's has already started to study the security and privacy protection of edge computing, a frontier and key research topic.

2.3.3   Gradual Improvement in China's Speed and Capability of Technological Innovation in Big Data

As information technology penetrates into human production and life, global data has seen explosive growth and massive agglomeration. The application scope of big data technology has expanded from traditional telecommunication and financial industries to industrial, healthcare, education and other fields. With the gradual improvement of China's big data policy system, its unique large-scale use cases and multi-type practice models have boosted a rise in the speed and capability of technological innovation in big data.

2.3.3.1   Big Data Applications Continue to Expand

In July 2019, Alibaba released the Apsara big data platform, which could be expanded to a cluster of 100,000 physical servers. According to relevant data, Apsara, as China's only self-developed big data cluster of computing engines, has surpassed those of Microsoft, Amazon and other major companies in size.[5]

In terms of big data applications, with abundant landing cases of platform, management and application technologies, fruitful research results have been produced on these technologies, which have produced many types of practice models. In May 2019, Baidu released the Dianshi big data platform to meet the future trend of data fusion and provide underlying data technologies and services. It also built a one-stop platform for online development, service hosting and transaction circulation driven by data security fusion. In July 2019, Huawei Cloud launched a series of Kunpeng big data services, including the MapReduce Service, Data Warehouse Service and Cloud Search Service, in order to accelerate the intelligent upgrading of enterprises. In August 2019, Huawei Cloud released BigData Pro, a Kunpeng

---

[5]"Alibaba Cloud's Apsara Big Data Platform: A Computing Platform with the World's Largest Clusters", *Chongqing Business Daily*, July 29, 2019.

big data solution that took Kunpeng computing power with unlimited expansion as the computing resource. It is a new public cloud big data solution characterized with "storage and computing separation, super elasticity and ultimate efficiency". In August 2019, Tencent Cloud released five new strategic products, the database intelligent housekeeper DBBrain, the cloud native database CynosDB, the cloud database TBase, the disaster recovery service DBS, and the Redis Hybrid Storage, to help millions of enterprises to deploy cloud services.

2.3.3.2 Big Data Products Are Highly Efficient

In November 2018, three of Alibaba Cloud's products, MaxCompute, DataWorks and AnalyticDB were selected into The Forrester WaveTM: Cloud Data Warehouse, Q4 2018 by Forrester Wave, a global authoritative advisory body. They ranked the seventh in the world and the first in China in terms of product capability, and surpassed Microsoft in Current Offering.[6] In May 2019, the Tencent Big Data Suite (TBDS) was listed in the Top 10 Big Data Cases and Top 100 Excellent Big Data Cases in China in the 2019 Big Data Expo. TBDS provides a one-stop data processing and mining platform for data access, analysis, governance and management in big data processing cases from TB to PB level.[7]

Meanwhile, it is also important to note that despite breakthroughs in some big data technologies in China, most of them are second innovation of foreign open source products. Relevant big data processing tools such as data acquisition, data processing, data analysis, data visualization technologies are essentially "stones from other hills" in a Chinese proverb. Innovation in systematic, platform-based core technologies remains rare. For example, self-developed technologies account for less than 10% of China's mainstream big data platform technologies.[8] Facing the future, China shall step up its efforts in making breakthroughs in independent core big data technologies.

2.3.4 Accelerated Development of VR Technology

By combining the technologies in multimedia, sensor, new display, Internet, AI and other fields, VR can greatly improve the mankind's perceptual ability, transform product forms and service models, and significantly impact on economy, science and technology, culture, military and people's life. At present, as various innovative bodies are marching into the VR field, technological progress and product upgrading

---

[6]Data Source: *The Forrester WaveTM: Cloud Data Warehouse*, Q4 2018.

[7]Data Source: *The Casebook of Excellent Products and Application Solutions of Big Data (2019)* released by the China International Big Data Industry Expo 2019.

[8]Prospective Industry Research Institute, "Market Analysis of China's Big Data Industry in 2019: Significant Progress and Four Suggestions to Address Five Development Challenges", March 14, 2019.

are accelerating, a wide variety of innovative applications are constantly emerging, and global leading enterprises are stepping up their VR deployment. As the core technologies in China's VR industry have further matured in recent years, China has become the world's most important manufacturer of VR terminal products. In the context of the commercial uses of 5G technology, the VR technology has become increasingly popular in 2019. In particular, progress has been made in the acquisition and modeling technology, analysis and utilization technology, exchange and distribution technology, display and interaction technology, among others.

2.3.4.1 Achievements are Made in the Acquisition and Modeling Technology

To acquire digital content, convenient and popular acquisition devices are necessary, and the content to be acquired shall be highly flexible and complete. For digital content generation, with foundation in the cartoon animation industry, its annual productivity of digital content reaches 10,000 min. An online independence-collaboration production line for digital content generation has been established. The two-dimensional animation production technology has matured and been applied. However, high quality 3D animation and re-usable basic material library remain in the development phase currently. A high-quality, highly efficient and complete technological service system for animation production and processing has not yet formed on the whole. For digital content creation, China's online game technology has been greatly improved and is closing the gap with the international advanced level. As Chinese enterprises have transformed their production model from agency operation to independent research and development, China has gradually scaled up its production of digital content creation tools. Companies like Shanda Networking and Kinsoft have developed a series of online games and game engines with proprietary intellectual property rights. Other independently developed game engines include Perfect World's Angelica and NetDragon's 2D and 2.5D products. But generally speaking, China's manufacturing of game engines started quite late and its game engine products were not complete.

For modeling technology, in the sub-field of surface attribute acquisition, many domestic companies are actively studying multi-dimensional, panoramic, dynamic lightfield acquisition technologies and systems. In another sub-field, motion capture system, China has developed its own low-cost and marker-based motion capture system.

2.3.4.2 Analysis and Utilization Technology, Exchange and Distribution Technology and Relevant Infrastructure Become Much More Important

Analysis and utilization technology is a key research topic supported by the National Natural Science Foundation of China, the 973 Program and the 863 Program. But the existing research focuses only on single machine algorithm and fails to sufficiently involve network and service aspects. Despite rapid development of some domestic companies, they imitate others and have few independently developed products. As the core of exchange and distribution technology, open content exchange and copyright management technology has relatively backward infrastructure. In particular, relevant resource management platforms, transaction platforms, transmission platforms and other service platforms for content production, release, circulation and consumption urgently necessary for content industry development have not been established yet. Due to the lack of open content exchange and transaction infrastructure, China's industrial chain of digital content has not really taken shape.

2.3.4.3 Display and Interaction Technology Is Developing Towards Human-Computer Interaction

In terms of display technology, remarkable progress has been made in China's naked eye 3D display devices and helmet-mounted displays. The 360-degree naked eye 3D display devices have fulfilled the breakthrough in research and development. At present, China's technology in this field has reached the level of similar foreign equipment, but efforts are still required to improve their reliability and industrialization level. For helmet-mounted displays, China has developed the principle prototype for the new-generation system of lightweight see-through head-mounted displays (HMD) with large field-of-views, superior to similar foreign products under patent protection in terms of technological indicators. For true 3D display technology, remarked progress has been made in high-speed projectors and spiral screens, rotating LED display screens, multi-layer LCD screens and varifocal electrowetting lens, but there remains a large gap in technological maturity with those in foreign countries. In June 2019, Beijing Ned and Xvisio Technology, a new computer vision company, unveiled a 6FOD MR head-worn display with super large field-of-views. It can interact with people through gesture and object recognition and meet the users' demands for smooth, jitter-free images.

In terms of human-computer interaction, progress has been made in interactions based on vision, tactile sense and sensors through efforts to pursue the trend of natural and harmonious interaction. In the field of vision-based interaction technology, the research mainly focuses on tracking, positioning and gesture recognition and processing, and progress has been made in tracking and positioning with and without identifiers. In June 2019, uSens released a 3D skeletal tracking and gesture recognition system running on a monocular color camera. It can recognize 26 degrees of freedom of 22 hand joints, including their 3D position and 3D rotation information. In terms of haptic interaction technology, the efforts are mainly concentrated on the theoretical research of haptic physiology and psychology experiments. The development of haptic devices stays basically in the stage of laboratory-based principle verification and has a large gap with foreign products. The human-computer interaction technology based on EMG sensors is quite creative, but relevant devices need to be further improved for easy use.

2.3.5 Sustained Efforts in the Research and Development of Quantum Information Technology

On the whole, China has some advantages in quantum communication research and technological accumulation, especially in the establishment of free-space quantum entanglement, teleportation and key distribution based on the Mozi satellite. The fiber-optic quantum key distribution technology based on the quantum government network in Wuhu and the Beijing-Shanghai Trunk Line has reached the world's top level. Nonetheless, China is still working hard to catch up with other countries in the development of quantum computing chips, quantum chemical simulation, quantum machine learning, quantum precision measurement and other fields.

Practical results have been achieved in quantum communications. The quantum key distribution (QKD) technology has been most fully developed. China's decoy-state SKD schemes has been formed, including the free-space QKD with a distance of 1,200 km based on the quantum communications satellite Mozi, and the fiber-optic SKD with a key generation rate of more than 20 kb/s based on the Beijing-Shanghai Trunk Line. The research team of Tsinghua University validated the security of the twin-field quantum key distribution (TF-QKD) without the help of quantum relay.[9] Other communication protocols are also growing mature and pushing forward the construction of a quantum network. In particular, the feasibility and security of the quantum secure direct communication (QSDC) have been validated, laying a strong foundation for the subsequent practical process.[10]

The research and deployment work of quantum computing has been in full swing. Alibaba, Tencent, Baidu, Huawei and other companies have actively deployed quantum computing. The Chinese Academy of Sciences, Tsinghua University, Zhejiang University, the University of Science and Technology of China, and other universities and research institutes have continued their research on quantum computing. Fruitful results have been produced in hardware implementation and software development. Some of them, such as quantum cloud platforms, can be applied into areas such as basic research and quantum computing education in the short term. Similar to international mainstream technological solutions, quantum computing in China is physically implemented mainly through superconducting solid-state systems, together with ion trap, liquid-state nuclear magnetic resonance, nitrogen-vacancy (NV) color center and other platforms. Huawei publicly released the HiQ cloud service platform for quantum computing simulator, which can simulate quantum circuits with 42 qubits for full-amplitude simulations, and 81 qubits for single amplitudes. It can perform error-correction simulations of circuits with tens of thousands of qubits. With a performance 5–15 times better than similar simulators, it provides great convenience for related research and education. Hefei Benyuan

---

[9]Source: *Physical Review X*, Issue 3, Volume 8, 2018.

[10]Source: *Light: Science & Applications*, Issue 1, Volume 8, 2019.

Quantum Computing Technology released a full set of quantum computing products, covering quantum chips, quantum measurement and control devices, quantum software and quantum cloud services.

Besides, as the measurement method and accuracy of quantum sensors in complex cases have further improved, they have promising applications in the high-precision measurement of atomic clocks, quantum gyroscopes and other devices. Researchers from University of Science and Technology of China first achieved Heisenberg limit measurement of noncommutative kinetic parameters, laying the foundation for subsequent research and development.[11]

At present, global scientific and technological innovation enters an unprecedented period of intense activity. Particularly, network information technology has accelerated breakthrough in innovation. In such a context, China is facing both development opportunities and risks and challenges. It should bolster areas of weakness and especially speed up addressing problems of key technologies in basic and frontier fields. As China's market advantages and application advantages have converted into a strong impetus to the development of network information technology, the breakthroughs in key and core technologies, new technologies, new applications and new forms of business are expected to accelerate innovative development, which has created favorable conditions for new economic and social development and reform.

---

[11] Source: *Physical Review Letters*, Issue 123, 2019.

Chapter 3
Development of Digital Economy

3.1 Outline

China is currently in a critical period of the transition from rapid growth to high-quality development. To vigorously develop digital economy is an important step to implement the principles of consolidating the gains made in the five priority tasks, strengthening the dynamism of micro entities, upgrading industrial chains, and ensuring unimpeded flows in the economy. It is of great significance for deepening the supply-side structural reform, promoting the shift in driving forces for development and achieving high-quality development. Against the backdrop of complicated and rigorous international situations and mounting downward pressure on the world economy, China's digital economy has expanded rapidly and its digital economy structure has been improved and optimized. Besides, industrial digitalization has advanced in depth and digital industrialization has developed steadily. In addition, digital economy's employment absorption capacity has significantly enhanced and the new growth drivers of economy have continued to expand.

Over the past year, steady progress was made in China's digital economy, an important driver of China's steady economic growth and development. In terms of total economic output, China's digital economy reached ¥ 31.3 trillion in 2018, a nominal growth of 20.9% when calculated in comparable terms, accounting for 34.8% of the country's total GDP,[1] a year-on-year growth of 1.9%. Its contribution toward GDP growth reached 67.9%, a year-on-year growth of 12.9%. Structurally, the value of digital industrialization in 2018 amounted to ¥ 6.4 trillion, indicating its entry into a period of steady growth, and the value of industrial digitization reached ¥ 24.9 trillion. The integration of digital economy and real economy has continued to deepen.

---

[1]Data Source: *The Digital China Construction and Development Report (2018)* released by the National Internet Information Office.

© Publishing House of Electronics Industry 2021
Chinese Academy of Cyberspace Studies, *China Internet Development Report 2019*, https://doi.org/10.1007/978-981-33-6930-6_3

The booming digital economy has a deep impact on employment and becomes a new channel for employment structure optimization and stable employment. New business models and forms of digital economy, such as online shopping, sharing economy and livestreaming, have created new patterns of flexible employment and spurred a sharp increase in the number of flexible employees.

3.2 New Engines Playing an Increasingly Prominent Role

At present, the rapid penetration and diffusion of emerging technologies has become a new driving force of economic growth, and information and communication technology (ICT) represented by big data, cloud computing, 5G and AI is boosting the rapid development of digital economy. At the same time, as the downward pressure on global economy continues to increase, the international competitive landscape becomes increasingly severe and complicated, and the share of Chinese labor-intensive products in the global market declines. It has become important to enhance the digitalization, networking and intelligentization level of China's real economy in order to strengthen its global competitiveness.

3.2.1 Digital Economy Creating Conditions for the Shift in Driving Forces for Development

Digital economy has not only greatly changed the form of goods and services, but also deeply impacted production factors such as the laborers, production tools, management level and operation model. It is increasingly becoming a new driving force of high-quality economic development. As China has strengthened its advantages over data size, data has become an important basic resource and has constantly been translated into value and efficiency. Driven by data flow, flows of technologies, materials, capital and human resources have converged and promoted the rapid growth of total factor productivity and the optimization of resource allocation. The construction of information resources in basic and key fields has been improved, and the advances in information and communication technology have further boosted innovation in the same field. Internet, big data and AI are accelerating the deep integration with real economy and its amplification, superposition and multiplying roles in economic development are unleashed gradually. In 2018, China's telecom market size reached ¥ 6.5556 trillion, an increase of 137.9% over 2017. Electronics and information manufacturing industry achieved a stable performance while at the same time securing progress. The added value of electronics and information manufacturing industry above designated size increased by 13.1% over the previous year, higher than the industrial growth above designated size, which was 6.9%. With the rapid innovation in Internet industry, new business forms such as mobile payment and cross-border e-commerce are constantly emerging and expanding, which are playing an increasingly important role in supporting economic and social development. In 2018, the business income of Internet and related industrial enterprises above designated size reached ¥ 956.2 billion, with a year-on-year growth of 20.3%.

3.2.2 Digital Economy Enhancing People's Sense of Happiness and Gain

General Secretary Xi Jinping stressed that the cyberspace affairs must uphold a people-centered development vision, with people's well-being as the starting point and footing for IT development. Currently, network information technology is integrated into people's life at an unprecedented speed scale and depth. As information service patterns are increasingly diverse and information products are updated rapidly, information consumption is shifting rapidly from low-level supply-demand balance to high-level supply-demand balance, and is playing an increasingly prominent role in improving people's livelihood and promoting innovation in social governance. New business models and forms such as "Internet + education", "Internet + medical care" and "Internet + travel" focus on keeping up with people's ever-growing needs for a better life and enhance people's sense of happiness and gain. Internet is utilized to promote targeted poverty reduction and elimination. Solid progress has been made in the five projects of Internet-assisted poverty alleviation, namely, the network coverage project, the rural e-commerce project, the network fostering intellectual project, the information services project, and the network public welfare project. Poverty alleviation projects have realized full coverage throughout China. The poor masses now have more approaches to become rich and more people are equally entitled to the dividend of technological changes.

3.2.3 Digital Economy Providing Impetus to Coping with the Downward Pressure on Global Economy

Global economy is now in the middle stage of a new round of Kondratieff Wave, in which traditional growth engines are providing less of a boost to economic growth and the risk of entering "new mediocre" is not yet eliminated. Policy changes of major economies and the resulting spillover effect bring uncertainties to the world. For example, the United States issued a series of macro policies on raising interest rates, reducing taxes and unwinding the balance sheet, which impacted the global market, shook the foundation of world economic recovery and hindered the process of global investment and trade recovery. According to the macroeconomic data released by the National Bureau of Statistics in July 2019, China's economy maintains steady growth and presents strong resilience and potential. Digital economy, as a "multiplication factor" to promote productivity growth, brings new sources of economic growth. Taking Chinese equipment manufacturing enterprise as an example, China is home to two of the world's top five network equipment manufacturing enterprises. Huawei has surpassed Ericsson AB to become the world's largest telecom equipment supplier. Huawei, Lenovo and Inspur are among the world's top five in terms of server shipments. BOE has jumped to the world's second largest large-size shipments panel factory. China is home to three of the world's top ten Internet companies in terms of global market value.

3.3 Gradual Improvement in China's Macro Policy Environment

The Central Economic Work Conference held in December 2018 stressed that China should utilize its new competitive advantages in technological innovation and scale effect to foster and develop new industrial clusters. According to *2019 Report on the Work of the Government*, China will strengthen the R&D and application of big data and AI technologies, foster clusters of emerging industries like next-generation information technology, high-end equipment, biomedicine, new-energy automobiles and new materials, and expand digital economy. As central departments accelerate the introduction of policies, local governments at all levels deepen their implementation and all regions give full play to their own advantages. A development pattern of digital economy with the features of linking up higher and lower levels, joint coordination and complementary advantages has formed.

3.3.1 Strengthening the Guiding Role of Strategic Plans

China has thoroughly implemented major strategies such as "Internet Plus" action plan, Outline of the National Informatization Development Strategy, and Action Plan on Promoting Big Data Development in order to support and promote the development of digital economy on top-level design. 19 departments including the National Development and Reform Commission (NDRC) jointly issued *Guiding Opinions on Developing Digital Economy to Stabilize and Expand Employment*, which set forth several measures in aspects such as accelerating the cultivation of emerging job opportunities in the digital economy field, continuously enhancing the digital competence of employees, and vigorously promoting the digital transformation of employment and entrepreneurship services. The Ministry of Industry and Information Technology (MIIT) and the State-owned Assets Supervision and Administration Commission of the State Council (SASAC) implemented the Special Action of Accelerating the Fostering of New Drives of Economic Growth to further improve the

Table 3.1   Policy documents on digital economy issued in 2019

Issuing time	Issuing agency	Title
February 2019	The Ministry of Industry and Information Technology of the People's Republic of China (MIIT), National Radio and Television Administration, China Media Group	Ultra HD Video Industry Development Action Plan (2019–2022)
May 2019	General Office of the Central Committee of the CPC, General Office of the State Council	Outline of Digital Countryside Development Strategy
July 2019	The Ministry of Transport	Outline of Digital Transport Development Plan
August 2019	General Office of the State Council	Guiding Opinions on Promoting the Well-regulated and Sound Development of the Platform Economy

supply capacity, strengthen the weakness and optimize the development environment for information and communication industry, as well as to promote the development of digital economy and the expansion and upgrading of information consumption. The General Office of the Central Committee of the CPC and the General Office of the State Council issued *Outline of Digital Countryside Development Strategy*. The outline proposed to focus on exerting the diffusing effect of information technology innovation, the spillover effect of information and knowledge and the inclusive effect of digital technology, bridge the "digital divide" between urban and rural areas, accelerate the modernization of agriculture and rural areas, and promote the modernization of rural governance system and capacity. Table 3.1 lists some of the policy documents on digital economy issued in 2019.

3.3.2   Accelerating the Implementation of Relevant Policies

Local governments at all levels have taken digital economy development as an important measure to promote high-quality economic development, and accelerated the implementation of relevant polices. Zhejiang Province prioritized digital economy as its "No. 1 project", vigorously developed the new economy with digital economy as the core and accelerated the construction of a modern economic system. Fujian Province further implemented the *Opinions of the General Office of Fujian Provincial People's Government on Accelerating the Innovation and Development of Industrial Digital Economy in Fujian Province*. It acted on the three principles of innovation driving and integrated development, market dominance and government guidance, and prudential tolerance and security specification and promoted the penetration of digital technology into all fields and links of industry in order to boost vigorous industrial development. Henan Province issued *Work Priorities of Henan Province's*

Digital Economy Development in 2019, which proposed to accelerate the construction of a new ecosystem of digital economy and build a national first-class big data industrial center, a cluster region for the development of emerging digital industries, and a national pilot region for digital economy development. Guangxi Zhuang Autonomous Region further implemented *Digital Economy Plan for 2018–2025 in Guangxi*. According to the plan, Guangxi actively promoted the agglomerational development of digital economy, and focused on fostering and developing digital industries like big data, cloud computing, AI, IoT, blockchain, integrated circuits, manufacturing of Intelligent terminals, software and information technology, and the BeiDou Navigation Satellite (BDS). Shaanxi Province issued *Work Priorities of Promoting the Development of "Three Economies" in Shaanxi Province in 2019*, which proposed to promote digital economy development through implementing the digital countryside construction project and building provincial-level digital economy demonstration parks and bases in a coordinated and standardized manner. Tianjin issued *Action Plan of Tianjin Municipality for Driving Growth in Digital Economy (2019–2023)*, which proposed the goal of building Binhai New District into a national digital economy demonstration zone by 2023. Table 3.2 lists some policy documents on digital economy issued by some provinces (municipalities directly under the Central Government) in 2019.

Table 3.2 Policy documents on digital economy issued by some provinces (Municipalities Directly Under the Central Government) in 2019

Province (municipality directly under the Central Government)	Issuing Time	Title
Fujian Province	January 2019	Notice on Issuing the Action Plan of "Digital Fujian, Broadband Project" in the New Era
	March 2019	Notice on Issuing the Work Priorities of the "Digital Fujian" Project in 2019
Shandong Province	March 2019	Notice on Issuing the 2019 Digital Shandong Action Plan
Tianjin City	May 2019	Action Plan of Tianjin Municipality for Driving Growth in Digital Economy (2019–2023)
Shaanxi Province	May 2019	Work Priorities of Promoting the Development of "Three Economies" in Shaanxi Province in 2019
Heilongjiang Province	June 2019	Digital Longjiang Development Plan (2019–2025)
Henan Province	July 2019	Work Priorities of Henan Province's Digital Economy Development in 2019

3.3.3   Accelerating Coordinated Regional Development

3.3.3.1   The Beijing–Tianjin–Hebei Area Promotes the Integrated Development of Big Data Industry

Beijing, Tianjin and Hebei Province signed a cooperation agreement on the coordinated development of informatization and jointly built the first national cross-regional big data comprehensive experimental zone to give full play to the comparative advantages of each region and facilitate data access, utilization and sharing in more industries and enterprises. Through coordinated industrial development, Xiong'an New Area became the destination of relocation of innovative and high-growth technology enterprises in Beijing, thus helping optimize the pattern of industrial chain collaboration. In 2018, digital economy in Xiong'an New Area accounted for more than 40% of its GDP, with the growth rate exceeding 14%.

3.3.3.2   The Yangtze River Delta Region Builds a Digital Economy Industrial Cluster

The Yangtze River Delta region further implemented *Three-Year Action Plan for the Integrated Development of the Yangtze River Delta Region (2018–2020)*. According to the plan, the National Integrated Circuit Innovation Center and the National Intelligent Sensor Innovation Center were established, and livelihood projects were jointly carried out and their results were shared across the region. Other efforts included promoting the mutual sharing and exchange of shipping logistics information and focusing on building the area into a highland for the development of global digital economy and a world-class city cluster with global competitiveness. Digital economy of the Yangtze River Delta region totaled 8.63 trillion *yuan* in 2018, ranking the first in terms of the size of digital economy in the country.

3.3.3.3   The Guangdong-Hong Kong-Macao Greater Bay Area Seizes the Commanding Point in Digital Economy Construction

The Greater Bay Area promoted the integrated and innovative development of digital economy, explored the construction of smart city clusters and deepened the pilot and demonstration projects of new smart cities. In specific, the Pearl River Delta region utilized the aggregation effect of the National Supercomputing Centers in Guangzhou and Shenzhen as well as the advantages of information communication industry in Guangzhou, Shenzhen and other areas to promote the construction of big data comprehensive experimental zones and accelerate the pace of breaking the development bottleneck of intelligent manufacturing. In 2018, digital economy in the Pearl River Delta region accounted for 44.3% of its GDP, the highest among all regions in China.

3.3.3.4 Northwest China Promotes Economic Revitalization by Developing Digital Economy

In 2018, Northwest China saw a faster growth of digital economy than the Beijing–Tianjin–Hebei Area and the northeast old industrial base, reaching 16.7%. Yet, the internal structure of its digital economy still needed improvement. In this region, industrial digitalization accounted for 90.8% of digital economy. Thanks to digital economy, some provincial capitals in Northwest China such as Xi'an have gradually narrowed the gap with eastern and central regions. Digital economy has created conditions for their rise to new first-tier cities.

3.3.3.5 Digital Economy Becomes a New Growth Point of the Comprehensive Revitalization of Northeast China

In Northeast China, efforts are made in developing digital economy in old industrial areas to promote the shift in driving forces for development in the old industrial bases and press ahead with the intelligent manufacturing of key industries. In 2018, the old industrial bases in Northeast China reported 11.3% growth of digital economy. Nevertheless, digital economy only accounted for 25.6% of the GDP. The proportion of digital industrialization is quite large, and the integrated industrial development remains to be strengthened, with a large space for digital economy development. Shenyang rolled out an intelligent upgrading project and built the largest robot industrialization base in China, which accelerated the digital transformation of the traditional heavy industrial city.

3.4 Industrial Digitalization Developing Towards Deep Integration

New technologies are rapidly penetrating into all areas of real economy. In particular, digital technology is reshaping the business forms, manufacturing processes, management models and development concepts of traditional industries, promoting the deep reconstruction and reform of information chain, industrial chain and value chain, and comprehensively improving the production efficiency of traditional industries through boosting accurate production and operation.

3.4.1 Industrial Internet Promoting the Digital Transformation of Enterprises

In 2018, China's digital economy accounted for 18.3% of GDP, with a year-on-year increase of 1.09%. It lied between the proportions of the service sector and agriculture sector and showed a tendency of accelerated growth. In traditional industries such as petroleum and gas extraction, ferrous metal mining, textile and garment, furniture, pharmaceuticals, steel/iron and their castings, automobile and household appliances, the digital economy share of GDP in 2018 all increased by about 1% over 2017.

3.4.1.1 Major Breakthroughs Are Achieved in Industrial Internet

On the supply side, China's network support capacity has been greatly enhanced, and the five national top-level nodes in the "east, west, south, north and center" of the identification and resolution system have been initially established. The supply capacity of industrial Internet platforms continues to be strengthened. China now has more than 50 influential industrial Internet platforms, each of which connects with an average of 590,000 devices. Active innovations have also been made in industrial apps. Besides, China has accelerated the construction of the security assurance system, systematically built national-level, province-level and enterprise-level security monitoring platforms, and sped up the popularization and application of independently developed security products.

On the demand side, remarkable achievements have been made in cost reduction and quality and efficiency improvement. Pilot enterprises have seen more than 20% growth in their labor productivity, and the energy consumption per ¥ 10,000 of industrial output value has reduced by more than 6%. Industrial innovation has accelerated. Chinese manufacturing enterprises have realized service transformation and stepped up their migration to the high-end part of the value chain by utilizing industrial Internet. In addition, interconnected development has produced abundant results, the capacity of resource convergence throughout the industry has been constantly enhanced, and cross-industrial and cross-regional enterprise cooperation and industrial agglomeration have further deepened. For example, Rootcloud connects more than 470,000 industrial devices on its industrial Internet platform. Goldwind, a wind power company, can monitor and operate nearly 19,000 wind turbines around the world online. It can also predict faults and carry out precautionary maintenance.

3.4.1.2 Industrial Enterprises Make Useful Attempts of Digital Transformation

Enterprises in discrete industries such as automobile, aviation and electronics have carried out diversified attempts of digital transformation. On the one hand, they have effectively integrated global design, manufacturing, service and intellectual resources on networking platforms to greatly shorten the development cycle of their products. On the other hand, they have established digital workshops and intelligent factories that vertically integrate the systems for field equipment production, production management and enterprise decision-making, which has enhanced the production flexibility and efficiency. For example, COMAC has established a global networked collaborative R&D platform for domestic cross-regional collaborative R&D and manufacturing, with which the development cycle of C919 aircraft has shortened by 20%, its production efficiency has increased by 30%, the manufacturing costs has reduced by 20%, and the incidence of manufacturing quality problems has decreased by 25%. Enterprises in process industries such as smelting and petrochemistry have also carried out comprehensive and systematic explorations in digital transformation. Leading enterprises have established comprehensive energy monitoring systems that cover the whole process ranging from energy supply, production, transfer, transmission to consumption, and built models for production and energy consumption prediction and capacity optimization. As a result, energy production and consumption have achieved integrated optimization and coordination, and the efficiency of energy production has been improved. For example, Sinopec Jiujiang builds an integrated energy management and control center as well as the model and optimization system for hydrogen and gas production-consumption balance with high value-added energy, which conducts closed-loop control on energy planning, energy production, energy optimization and energy evaluation, thus realizing the goal of energy conservation and consumption reduction. Its energy efficiency has increased by 4% over the past three years.

3.4.2 Deepening Digital Transformation in the Service Sector

The service sector has accelerated digital transformation and new business models and forms have emerged in various fields. In 2018, digital economy accounted for 35.9% of GDP in the service sector, up 3.28% over 2017, far higher than the industrial average. Among industrial divisions, the insurance, and radio, television, film video and radio production have the largest proportion of digital economy, reaching 56.4% and 55.5% respectively.

New business forms and models of digital services are constantly emerging, online retail and mobile payment are accelerating development, and the space for digital transformation of the service sector is constantly expanding. The service sector shows formidable leading advantages in spurring consumption growth, entrepreneurship and employment and industrial transformation.

3.4.2.1 Online Shopping

As of June 2019, the user size of online shopping was 639 million or 74.8% of China's total netizen population, up 28.71 million over the end of 2018; the number of mobile

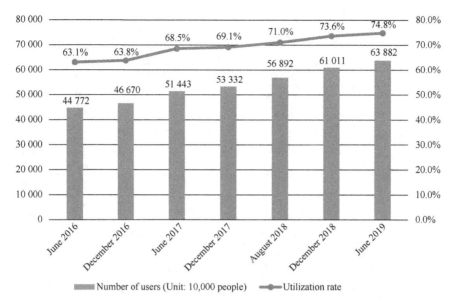

Fig. 3.1 User size and utilization ratio of online shopping in China from June 2016 to June 2019. *Data Source* The 44th "Statistical Report on China Internet Development"

shopping users had reached 622 million, up 29.89 million from the end of 2018, taking up 73.4% of mobile Internet users. In the first half of 2019, the online shopping market maintained rapid development, while the small-town and rural markets, cross-border e-commerce, and model innovation provided new growth momentum for the online shopping market. Figure 3.1 shows the user size and utilization ratio of online shopping in China from June 2016 to June 2019.

3.4.2.2 Online Meal Ordering

Up to June 2019, the user size of online meal ordering was 421 million or 49.3% of China's total netizen population, up 15.16 million over the end of 2018; the number of mobile meal ordering users had reached 417 million, up 20.37 million from the end of 2018, representing 49.3% of mobile Internet users. The online meal ordering business, as the foundation of the life service system, is deeply integrating itself with new retail and other related business, connecting more life service scenarios and boosting the coordinated development of the ecosystem. Figure 3.2 shows the user size and utilization ratio of online meal ordering in China from June 2016 to June 2019.

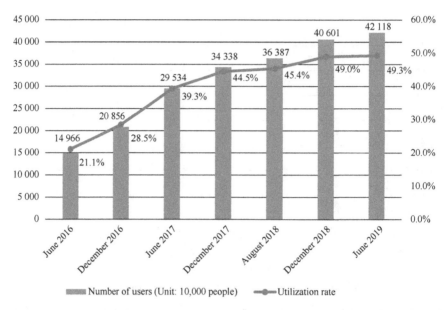

Fig. 3.2   User size and utilization ratio of online meal Ordering in China from June 2016 to June 2019. *Data Source* The 44th "Statistical Report on China Internet Development"

3.4.2.3   Online Travel Booking

As of June 2019, the number of online travel booking users in China had reached 418 million, up 8.14 million from the end of 2018, accounting for 48.9% of all Internet users. For the online booking of tourism and vacation products, different product combinations are produced based on users' subdivided needs, including family tour products suitable for children and the elderly and customized services for users' personalized needs. Besides, offline demands are integrated to conduct intelligent marketing through big data technology so as to tap into potential user markets. Figure 3.3 shows the user size and utilization ratio of online travel booking in China from June 2016 to June 2019.

As of June 2019, the user size of online car-hailing services was 337 million or 39.4% of China's total netizen population, up 6.7 million over the end of 2018; the number of online premier and fast ride had reached 339 million, up 6.33 million from the end of 2018, representing 39.7% of China's Internet users. Regarding policy supervision, the compliance of the online car-hailing industry achieved initial results. Online Car-hailing Business Certificate, Online Car-hailing Driver Certificate, and Online Car-hailing Transportation Certificate are the current market access conditions for online car-hailing services in China. As of February 2019, 247 cities across the country had issued specific suggestions and measures to regulate the development of online car-hailing services. More than 110 online car-hailing companies had obtained business licenses, and 680,000 driver licenses and 450,000 vehicle transportation licenses had been issued. The steady progress in the compliance of the

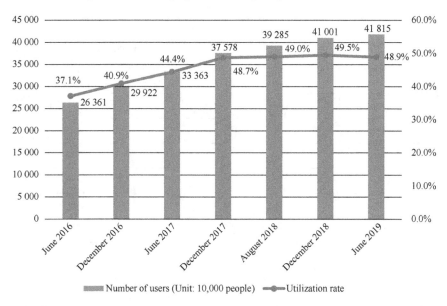

Fig. 3.3 User Size and Utilization Ratio of Online Travel Booking in China from June 2016 to June 2019. *Data Source* The 44th "Statistical Report on China Internet Development"

online car-hailing industry has resulted in an increasingly standardized operation and a fairer and more orderly competitive environment. Figure 3.4 shows the user size and utilization ratio of online car-hailing services in China from June 2016 to June 2019.

3.4.2.4 Financial Technology

As of June 2019, the user size of Internet wealth management in China had reached 170 million, up 18.35 million from the end of 2018, accounting for 19.9% of all Internet users. The user size of online payment was 633 million or 74.1% of China's total netizen population, up 32.65 million over the end of 2018.

In recent years, with continuous efforts in supervision and compliance, the chaos in Internet financial industry has been further normalized. Policies such as *Guiding Opinions on Promoting the Healthy Development of Internet Finance* and *Implementation Plan for the Special Campaign against Internet Financial Risks* have been effectively implemented. Other documents such as *Guidelines for the Online Lending Fund Depository Business* and *Guidelines for the Disclosure of Information on the Business Activities of Online Lending Information Intermediary Institutions* have been released successively. The archival filing work of Internet finance has been launched gradually. In the first half of 2019, Internet financial industry continuously improved its overall standardization level and showed an upward trend of development. On the one hand, financial investment decisions were made in a more

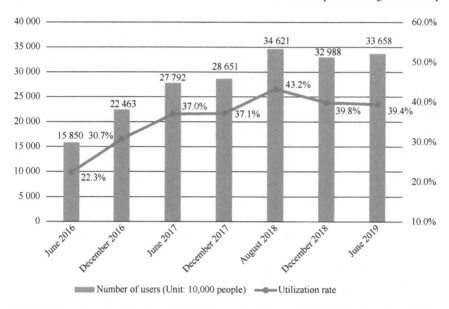

Fig. 3.4 User size and utilization ratio of online car-hailing services in China from June 2016 to June 2019. *Data Source* The 44th "Statistical Report on China Internet Development"

intelligent manner. Financial institutions utilized big data and cloud computing technologies to systematically analyze borrowers' detailed analysis data and develop exclusive investment portfolios for investors. On the other hand, the coverage of inclusive financial services expanded. Financial technology was employed to evaluate the credit of small and micro businesses and make inclusive financial services available to them. For example, Ningbo Central Branch of the People's Bank of China built an inclusive finance credit information service platform to provide credit approval and risk management tools for 64 banks, micro-credit companies and insurance companies. More than 7,000 inquires were made on the platform every day. MYBank utilized big data and other technologies to solve the financing difficulties of micro businesses on e-commerce platforms that had no secured loans, credit records or financial statements. It provided services to about 11 million small and micro businesses. Figure 3.5 shows the user size and utilization ratio of Internet wealth management in China from June 2016 to June 2019.

3.4.2.5   Online Education

As of June 2019, the user size of online education was 232 million or 27.2% of China's total netizen population, up 31.22 million over the end of 2018; the number of mobile learning users had reached 199 million, up 5.3 million from the end of 2018, accounting for 23.6% of mobile Internet users. Figure 3.6 shows the user size and utilization ratio of online education in China from June 2016 to June 2019.

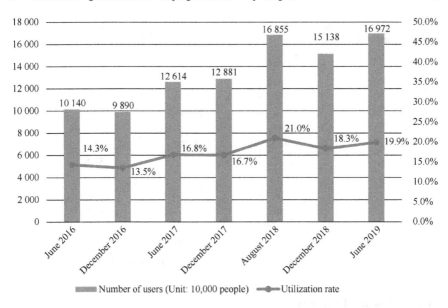

Fig. 3.5 User size and utilization ratio of Internet Wealth Management in China from June 2016 to June 2019. *Data Source* The 44th "Statistical Report on China Internet Development"

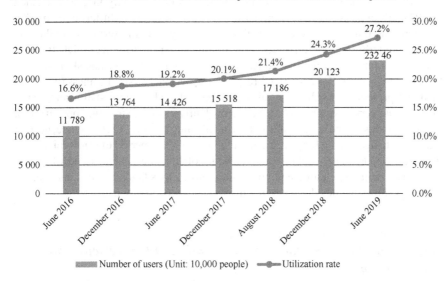

Fig. 3.6 User size and utilization ratio of online education in China from June 2016 to June 2019. *Data Source* The 44th "Statistical Report on China Internet Development"

Area	Size (100 million yuan)	Growth rate (%)
Transportation	2 478	23.3
Shared accommodation	165	37.5
Knowledge and skills	2353	70.3
Life services	15 894	23
Shared medical care	88	57.1
Co-working	206	87.3
Production capacity	8 236	97.5
Total	29 420	41.6

Table 3.3 Development of China's sharing economy in 2018

(*Data Source* National Information Center)

3.4.2.6   Sharing Economy

In 2018, the market transaction volume of the sharing economy in China reached ¥ 2,942 billion, which was an increase of 41.6% over 2017. In terms of market structure, the transaction volume of digital economy in living services, production capacity and transportation fields accounted for a larger proportion of GDP. In terms of the development speed, the production capacity and knowledge and skills saw the largest growth, up 97.5% and 70.3% respectively over 2017. New business forms and models represented by online car-hailing services, shared accommodation and shared medical care services have become new growth drivers of the structural optimization of service industry and the transformation of spending patterns. On the whole, despite dramatic changes in the bicycle sharing market in a short period of time which have triggered disputes and doubts from all walks of life, the general trend of digital economy's rapid penetration and integration into various fields remains unchanged. Table 3.3 shows the development of China's sharing economy in 2018.

3.4.3   *Continuous Digital Transformation in the Agricultural Sector*

3.4.3.1   Ubiquitous Network Access Boosts Rural Revitalization

In recent years, China has vigorously promoted Internet construction in rural areas. Policy documents such as *Opinions on the Implementation of the Rural Revitalization Strategy* and *Strategy Plan for Rural Vitalization (2018–2022)* have been released. In May 2019, the General Office of the Central Committee of the CPC and the

General Office of the State Council jointly issued *Outline of Digital Countryside Development Strategy*, highlighting that digital countryside was a strategic direction of rural revitalization as well as an important task of building "Digital China". China has now initially built an integrated, ubiquitous, safe and green broadband network environment and basically realized "city fiber-into-floor/home and rural broadband-into-township/village". As of June 2019, China's FTTH/O users reached 396 million, with the penetration rates of fiber optic and 4G networks both exceeding 98% in administrative villages[2]; the size of rural Internet users was 225 million or 26.3% of China's total netizen population, up 3.05 million over the end of 2018, with a half year growth rate of 1.4%.

3.4.3.2 Rural Areas Maintains Steady Growth in Digital Economy

Intelligent agriculture utilizes IoT technology to control agricultural production and further realize digitalization, networking and intelligentization of agricultural products. For example, Bairin Right (Balin You) Banner in Inner Mongolia Autonomous Region established an intelligent irrigation system to accurately predict the watering timing and quantity based on temperature, light, humidity, transpiration conditions and other data, which saved 1.5 times as much as the water volume of Hangzhou West Lake every year. The system solved the problem of groundwater recession caused by pumping irrigation and created a new model of sustainable agricultural development in dry areas. In 2018, digital economy accounted for 7.3% of GDP in the agricultural sector, up 0.72% over 2017. The digitalization level of agricultural production has been enhanced year by year and still has great space for improvement. The forestry sector has the highest share of digital economy in China, accounting for about 13%, which is followed by fishery, agriculture and animal husbandry, with the lowest being less than 5%, far lower than the average level of the service and industry sectors. It is predicted that by 2020, the potential market size of intelligent agriculture in China will grow to 200 billion *yuan* from 100 billion *yuan* in 2015, indicating tremendous market prospects.

3.4.3.3 The Quality and Safety Traceability of Agricultural Products Develop Towards Digitalization

China has officially launched the national MIS platform for quality and safety traceability of agricultural products. The platform allows for the centralized management of various information on product traceability, supervision, monitoring and law enforcement, and provides a unified query entry for fast, real-time inquiry of the traceability information of agricultural products. With its open and compatible features, the whole process of agricultural products "from farm to table" can be

---

[2]Data Source: The Ministry of Industry and Information Technology of the People's Republic of China (MIIT).

traced, thus effectively improving the efficiency of agricultural product quality and safety supervision. All provinces and cities in China have actively employed IoT technology and device to collect the information of agricultural products throughout the traceability chain, including logistics, information flow and people flow. On such basis, big data mining and analytical technology is utilized for effective supervision of the industrial chain of agricultural products.

3.4.3.4 Information Services Become More Popularized in Rural Areas

Mobile Internet is a key starting point to implement the information-entering-village project. Internet enterprises, industrial associations and professional institutions have increased investment in mobile application platform and content such as agriculture-related WeChat official accounts, Weibo accounts and professional applications to provide farmers with convenient information services of policy, market, science and technology and insurance in all aspects of production and life. They have also ensured timely transmission of information on agricultural policies and regulations, new varieties and new technologies, animal and plant diseases, agricultural product prices, and agricultural product quality and safety, cultivated the farmers' habits of utilizing Internet and increased penetration rate of Internet in rural areas.

3.4.3.5 Informatization Contributes to Targeted Poverty Alleviation

The Office of the Central Cyberspace Affairs Commission, the National Development and Reform Commission (NDRC), the Leading Group Office of Poverty Alleviation and Development of the State Council and the Ministry of Industry and Information Technology jointly formulated and issued *Work Priorities of Internet-Assisted Poverty Alleviation in 2019*, which focused on addressing the outstanding issues of "two no worries" and "three guarantees" for severely impoverished areas, specially disadvantaged groups and registered poor households, and fully tapped the potential of Internet and informatization in targeted poverty alleviation. Rural e-commerce companies have expanded rural information services on network platforms and built them into information service platforms for agriculture, rural areas, and farmers that cover all counties, townships and villages. According to statistics, in 2018, China had more than 9.8 million rural e-commerce enterprises, with the online retail sales of agricultural products reaching 230.5 billion *yuan*, a year-on-year increase of 33.8%.

3.5 Steady Progress in Digital Industrialization

At present, with the acceleration of innovation and breakthroughs in the new-generation information and communication technology represented by 5G, IoT, cloud computing, big data, AI and blockchain, China has stepped into the golden period of its full penetration into economic and social areas. With the focus of development shifting from human interconnection to IoT, from massive data to AI, from consumption upgrading to production transformation, digital economy will continue to go towards new heights.

3.5.1 Telecommunications Industry Playing a Greater Supporting Role

In 2018, China focused on improving its infrastructure capability and steady progress was made in the development of telecommunications industry. This played an increasingly important role in supporting national economic and social development. The total business volume of telecommunication services has maintained rapid growth. In 2018, China's telecom market size reached ¥ 6,555.6 billion (calculated at 2015 constant price), an increase of 137.9% over 2017. China's total telecommunications revenue in 2018 reached ¥ 1,301 billion, up 3% over 2017; the total revenue for the first half of 2019 reached ¥ 672.1 billion, roughly equivalent to that of the same period in the previous year. Mobile data traffic consumption has continued to grow rapidly. In 2018, mobile Internet traffic consumption reached 71.11 billion GB, a 189.1% increase compared to 2017, and the growth rate increased by 26.9% over 2017. The average monthly household traffic (DOU) of mobile Internet access in 2018 reached 4.42 GB/month/user, which was 2.6 times that of the previous year. Figure 3.7 shows the growth of China's total telecom business and revenue from 2010 to 2018.

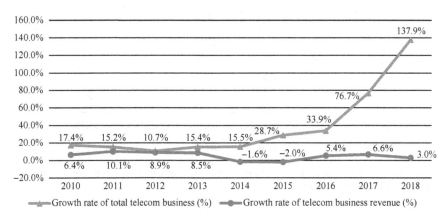

Fig. 3.7 Growth of China's total telecom business and revenue from 2010 to 2018. *Data Source* The Ministry of Industry and Information Technology of the People's Republic of China (MIIT)

3.5.2   Growing Business Income of Internet Enterprises

Chinese Internet enterprises have maintained a rapid development momentum. In 2018, the business income of China's Internet and related services enterprises (collectively referred to as "Internet enterprises") above the designated size reached ¥ 956.2 billion, a 20.3% increase over 2017. Internet business income in major provinces maintained a rapid growth in major enterprises, and Guangdong, Shanghai and Beijing were the top three, with their Internet business income increasing by 26.5%, 20% and 25.2% respectively. In the first half of 2019, the total business income of Internet enterprises above the designated size reached ¥ 540.9 billion, a year-on-year growth of 17.9%. In niche markets, the income from information services such as online music and video, online games, news and information, and Internet reading reached ¥ 370.3 billion or 68.5% of the total Internet business income in the first half of 2019, a 23% growth over the same period of the previous year, which increased by 5.8% compared with the first quarter. Figure 3.8 shows the growth in the business income of China's listed Internet enterprises, and Table 3.4 lists the top ten listed Internet enterprises in terms of market value.

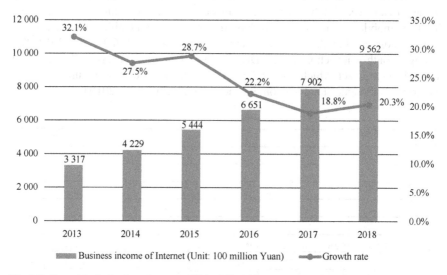

Fig. 3.8   Growth in the business income of China's listed internet enterprises. *Data Source* Financial statements of listed companies in China

Ranking	Corporate name	Market value (100 million US dollars)
1	Alibaba	4 499
2	Tencent	3 951
3	Meituan-Dianping	551
4	JD.com	453
5	Pinduoduo	391
6	Baidu	365
7	Netease	325
8	Xiaomi	261
9	Tencent Music	217
10	360	212

Table 3.4 Top Ten Listed Chinese Internet Enterprises by Market Value (As of August 2019)

3.5.3 Electronics and Information Manufacturing Industry Boosting Industrial Growth

China's electronic information manufacturing industry has developed rapidly, and its status in industry has been gradually upgraded. In 2018, the added value of the electronics and information manufacturing industry above designated size increased by 13.1% over the previous year, higher than the industrial growth above designated size, which was 6.9%; in the first half of 2019, the added value grew by 9.6% over the same period of 2018. In terms of shipments, the export delivery value of the electronics and information manufacturing industry above designated size in 2018 increased by 9.8% year on year, and that in the first half of 2019 grew by 3.8% over the same period of 2018. In terms of operating income and profit, the operating revenue of the electronics and information manufacturing industry in 2018 increased by 9.1% year on year, and that in the first half of 2019 grew by 6.2% over the same period of 2018. Figure 3.9 shows the changes in the growth in the total operating income and profit of the electronics and information manufacturing industry since June 2018.

3.5.4 Software and Information Technology Service Industry Witnessing Rapid Growth in Income

China has continuously adjusted and optimized the structure of the software and information technology service industry, and new growth points have been constantly emerging. On the whole, the total software business income of the electronics and information manufacturing industry in China in the first half of 2019 reached ¥ 3,283.6 billion, a 15% increase over the same period of 2019, and the growth rate increased by 0.6% year on year; the total profit in the first half of 2019 reached ¥

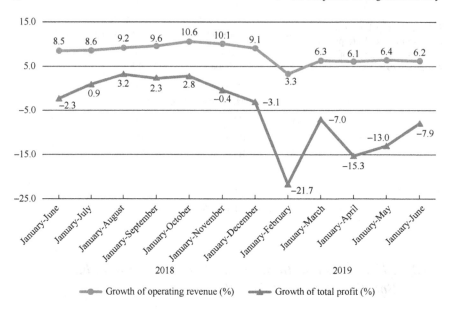

Fig. 3.9 Changes in the growth of the total operating income and profit of the Electronics and Information Manufacturing Industry Since June 2018. *Data Source* The Ministry of Industry and Information Technology of the People's Republic of China (MIIT)

408.8 billion, with a year-on-year growth of 9.9%. In terms of market segments, the income of software products, especially industrial software, grew at a higher speed. The business revenue of software products reached ¥ 918.3 billion or 28% of the total in the industry in the first half of 2019, up 14.1% year on year, and the growth rate increased by 0.5%. Industrial software products reached a business income of ¥ 84.4 billion, up 19.8% year on year, 4.8% higher than the average growth rate of the whole industry.

The income of information technology services, especially e-commerce platform technology services and big data services, realized rapid growth. In the first half of 2019, information technology services reached the business revenue of ¥ 1,938.6 billionor 59% of the total in the industry, up 17.2% year on year, with the growth rate increasing by 0.1% over the same period of 2018. In this field, the business revenues of cloud services and big data services increased by 14.6% and 20.5% respectively, which together accounted for 9% of the total revenue of information technology services; the business revenue of e-commerce platform technology services increased by 22.6%, up 7.6% year on year; the business revenue of information security products and services maintained double-digit growth. In the first half of 2019, information security products and services reached the business revenue of ¥ 50.09 billion, up 10.8% year on year.

3.6 Profound Impact on Employment

Digital economy has become a new channel for employment structure optimization and stable employment. New business models and forms of digital economy, such as online shopping, sharing economy and livestreaming, have created new patterns of flexible employment, offered more jobs for urban labor and promoted employment transfer of surplus labor in agriculture. However, they have also increased the risk of structural unemployment.

3.6.1 Digital Economy Creating New Patterns of Flexible Employment

With the vigorous development of digital economy, the number and proportion of flexible employees in China both increase rapidly. They are increasingly becoming an important part of China's overall employment, rather than just playing a supplementary role. It also features the expansion of areas of flexible employment and the transition from low-level employment to high-level employment and from passive choice to initiative participation. Take sharing economy as an example. In 2018, about 75 million service providers in China participated in sharing economy, most of which were part-time workers. 6.7% of Didi's car-hailing drivers were registered poor people, and 670,000 delivery riders from state-registered poor counties earned income through Meituan.[3]

Preliminary estimates suggested some 191 million jobs were offered in digital economy in 2018, accounting for 24.6% of overall employment. In April 2019, the General Offices of the Ministry of Human Resources and Social Security, the State Administration for Market Regulation and the National Bureau of Statistics jointly issued *Notice Regarding Releasing Occupational Information Including Intelligent Manufacturing Engineering and Technical Personnel and Other Occupations*, which confirmed information on 13 new occupations. Among them, AI engineering technician, IoT engineering technician, big data engineering technician and digital manager were all closely related to digital economy. Driven by digital economy, more employment patterns are created, the threshold to employment is further lowered and people get jobs more easily. It has also played an active role in "keeping employment stable". Figure 3.10 shows the employment in China's digital economy.

---

[3]Data Source: The Sharing Economy Research Center of the National Information Center.

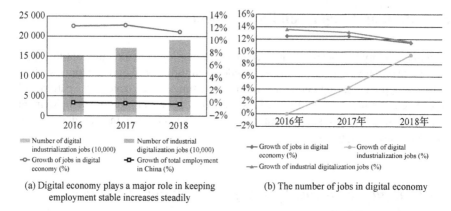

(a) Digital economy plays a major role in keeping   (b) The number of jobs in digital economy
    employment stable increases steadily

Fig. 3.10 Employment in China's digital economy. *Data Source* China Academy of Information and Communications Technology

3.6.2 The Risk of Structural Unemployment Still Needing to Be Guarded Against

With the rapid development of digital economy, digital technologies are widely applied into economic and social areas, which results in great changes in corporate productivity, organizational division of labor and industrial structure, and increases the risk of structural unemployment. Industrial robots are rapidly put into new posts of the production line, and service robots are becoming a new consumption object and a new hot spot in people's lives. China has more than 400,000 industrial robots, about one quarter of the world's total. Midea, Geely, Foxconn and other enterprises are stepping up their construction of automatic robot production lines. Unmanned aerial vehicles (UAV), unmanned ground vehicles (UGV) and humanoid robots are constantly applied into service fields such as logistics, food and beverage and medical care. For example, in Haidilao's 24 h unmanned hotpot restaurants, robots have been deployed to automate the whole service process including washing, dispensing and serving dishes. On November 11, 2018, Ali Xiaomi, Taobao's smart customer service chatbot, undertook 98% of online service requests, equivalent to the workload of 100,000 people of traditional customer services. China's manufacturing industry remains in the middle and low end of the global value chain, with its major business on production, assembly and other conventional services. The employees in the industry are highly replaceable, and if they are replaced, great employment pressure will arise accordingly. In addition, new business models and forms are constantly emerging. The shift from old to new business forms has created more jobs and caused a serious squeeze and impact on some traditional fields. Employees in some industries will face the risk of losing their jobs.

China's digital economy is characterized with huge development potential and strong driving effect. In the next step, continuous efforts are still needed to summarize the development experience and rules, give play to China's advantages as a major economy and market, maintain strategic focus, boost the dynamism of development and fully tap the potential of data resources. Other efforts include continuously fostering new forms of digital economy, accelerating digital transformation in economic and social areas, and promoting high-quality development by fostering new growth drivers.

Chapter 4
E-government Construction

4.1 Outline

To promote the modernization of the national governance system and governance capacity through informatization is important for the overall development of the cause of the Party and country as well as the fundamental interests of people. E-government is an important task of accelerating the construction of national informatization and furnishing the fruits of Internet development to hundreds of millions of people. It is also an important way to comprehensively deepen the reform of the administrative system and improve the government's ability to perform its duties, playing a fundamental and strategic role in the modernization of the national governance system and governance capacity.

Chinese government attaches great importance to e-government by focusing on following the law of e-government development and comprehensively advancing e-government construction. Over the past year, e-government has accelerated its development in China, with its governance system and mechanism improved constantly and major progress made in construction of "Internet + government services". Besides, efforts have been stepped up in the integration and sharing of government information system and the opening of public information resources, and the innovative application of the new-generation information technology in key areas has been continuously deepened. The basic support and guarantee system has been further strengthened, playing an active role in regulating the government's exercise of power, optimizing the supply of public services, stimulating market's innovation and vitality, and enhancing people's sense of gain.

© Publishing House of Electronics Industry 2021
Chinese Academy of Cyberspace Studies, *China Internet Development
Report 2019*, https://doi.org/10.1007/978-981-33-6930-6_4

4.2 Progressive Improvement in the Governance System and Mechanism of E-government

In recent years, with the comprehensive deepening of China's reform of its administrative system, the system of e-government construction as well as the governance system and mechanism have been gradually improved. A pattern of overall planning and coordination and joint administration and collective governance has basically taken shape.

4.2.1 Increasing Improvement in the Overall Planning and Coordination Mechanism

To promote unified e-government construction across the whole country and build a unified, efficient, convenient, open and interconnected e-government system is a necessary requirement of the high-starting point planning, high-standard construction and high-quality development of e-government in China in the new era. Chinese government has stepped up efforts in overall planning and coordination and focused on addressing issues in e-government development like sectoral fragmentation, multiple leadership, regional segmentation, multiple investment and lack of coordination to promote the healthy development of e-government.

4.2.1.1 Establishing and Improving the National Overall Planning and Coordination Mechanism for E-government

The national overall planning and coordination mechanism for e-government clarifies the functions and responsibilities of relevant central departments in e-government construction, management, operation and standardization and proposes solutions to major and important issues in e-government construction, aiming to improve the consistency and coordination of national e-government decision-making and deployment. The Office of the Central Cyberspace Affairs Commission organizes national conference on the overall planning and coordination of e-government annually to deploy annual e-government tasks. Member units of the national overall planning and coordination mechanism for e-government perform their own duties while strengthening coordination, resource integration and upper-lower linkage, thus improving the efficiency of e-government construction and management.

4.2.1.2 More Efforts Are Put in the Overall Planning and Coordination of E-government in Various Regions

On the one hand, provincial e-government comprehensive coordination mechanisms have been gradually established and the level of departmental e-government collaboration has been greatly improved. As of May 2019, 27 provincial-level administrative regions in China have set up e-government comprehensive coordination departments. On the other hand, cyberspace affairs departments have strengthened e-government development. As of May 2019, 28 provincial cyberspace affairs departments had been designated to fulfill e-government-related functions.

4.2.2 Accelerating the Establishment of the E-government Policy and Standard System

E-government construction is a systematic project featuring wide coverage, long construction period, higher technological requirements and complex business demands, and establishing a policy and standard system is a fundamental and key measure to advance this project. In recent years, in order to address the prominent issues in e-government construction, such as insufficient policies and inconsistent standards, Chinese government has accelerated the establishment of the e-government policy and standard system, which lays a sound foundation for e-government development.

4.2.2.1 Establishing the E-government Policy System

In recent years, China has released a series of policy documents, including *Guiding Opinions on Accelerating the "Internet + Government Services" Work*, *Notice on Issuing the Interim Measures for the Administration of Sharing of Government Information Resources*, *Notice on Issuing the Plan for the Integration and Sharing among Government Information Systems*, *Notice on Issuing the Guidelines to Develop Government Websites*, *Guiding Opinions on Further Promoting the Popularization of Examination and Approval Services*, and *Provisions of the State Council on Online Government Services*. Besides, other campaigns such as national government website survey, national e-government comprehensive pilot project, and government website intensification pilot project are organized and conducted, which have effectively sped up the progress of China's e-government development.

4.2.2.2   Accelerating the Progress of E-government Standardization

In recent years, the State Administration for Market Regulation (SAMR), the Standardization Administrative of China (SAC), and the Office of the Inter-Ministerial Joint Meeting for Electronic Document Management (State Cryptography Administration) jointly issued six national standards (including *Electronic Certificate*) to address issues such as difficulty in integrating and applying the certificate systems that were dispersedly established and managed, and the failure in wide-range mutual sharing and recognition of certificate information due to its inconsistency with documentary standards. These standards have provided support for building the national electronic certificate library and basic platform, realizing cross-level, inter-departmental and cross-regional mutual recognition and sharing of electronic certificate information, and promoting the sharing of government information resources related to certificates. The Ministry of Industry and Information Technology (MIIT) issued *Guidelines on the Construction of the Comprehensive Standardization System for Cloud Computing*, established standardization systems for infrastructure, services, resources, security and application, and formulated and released a series of standards for e-government cloud platforms. In addition, it has established information technology service standards to provide effective reference for the process management of government informatization construction.

4.2.3   Continuously Optimizing the Functions of E-government Management Departments

In 2018, the Chinese government carried out a new round of institutional reform that focused on improving the optimization, collaboration and efficiency of the Party and state institutions. Specific measures include transforming institutional organization, optimizing function allocation and building a service-oriented government able to satisfy the needs of people. Thanks to the reform, China's structuring of e-government management institutions has been constantly improved and their functions have been continuously optimized.

4.2.3.1   Establishing E-government Management Organizations

Guangdong Province has made innovation in and strengthened e-government management organizations. In 2018, the Government Affairs and Data Management Bureau were established to solve issues in e-government such as overlapping management and multiple leadership. The bureau integrates e-government, big data, government affairs and other functions that used to be undertaken by different departments and combines the permanent posts, personnel and strength of the information centers of different departments. As a result, issues such as separate construction

& maintenance and decentralized management have been effectively relieved, and Chinese government's governance capacity of overall planning and coordination and data resources integration has been greatly improved. Besides, other regions have also set up big data bureaus to press ahead with e-government construction. As of July 2019, 22 provincial administrative regions in China, including Tianjin, Anhui, Henan and Guangxi, had established their own big data institutions.

4.2.3.2 Joining Hands to Advance the Key E-government Work

All departments have been actively engaged in the project approval, management, acceptance and other processes of e-government projects. As of May 2019, the development and reform commissions in 26 provincial administrative regions, financial departments in 24 provincial administrative regions, and government offices in 11 provincial administrative regions had participated in the management of e-government projects; and 16 provincial administrative regions had realized integrated management of e-government funds.

4.2.4 Explorations Made in Comprehensive Pilot Projects

In 2018, in order to help local governments build e-government development models in line with local reality, produce fruitful referential achievements of e-government development, and accumulate experience for coordinated e-government development, the Office of the Central Cyberspace Affairs Commission, the National Development and Reform Commission (NDRC) and other relevant departments jointly launched the national e-government comprehensive pilot project in 8 provincial administrative regions with better basic conditions. The project focused on five aspects, namely, establishing a mechanism for coordinated development, improving the intensive level of infrastructure, facilitating the sharing of e-government information resources, promoting the "Internet + government services" work, and advancing the standardized application of electronic documents in key areas. After more than one year of hard work, the pilot regions have made positive progress in the national e-government comprehensive pilot work.

Beijing actively carried out the pilot work. It established government information resource sharing and exchange platforms at municipal and district levels that enabled inter-departmental and cross-level sharing of more than 1,300 types of data among 41 government departments. Besides, the e-government cloud for Beijing Municipal Government was built to connect with more than 800 information systems of more than 60 commission offices in Beijing.[1] Shanghai strengthened data interconnection on the online government affair hall, on which 1,274 service items of 39 municipal departments and 16 districts were handled uniformly. It achieved 100% service

---

[1] As of June 2018.

acceptance and handled a total of 3.67 million cases.[2] Jiangsu Province promoted the connection and sharing of the basic data of individuals and legal persons, as well as the theme data on platforms for investment project approval, credit information and "Integrating Certificates into One", making the user real-name authentication and data exchange and sharing among different platforms possible. Zhejiang Province established an electronic certificate library to collect data of 153 common certificates, including ID cards, business licenses and real property ownership certificates, which satisfied over 80% of the data sharing demands across the province. Fujian Province strengthened joint contribution and sharing of computing resources, storage resources, operation and maintenance service support, security and other infrastructure, and actively mobilized social forces to participate in infrastructure construction. Guangdong Province continuously improved the system and mechanism of digital government reform across the province and forged ahead the reform in all aspects, with preliminary results achieved so far. Shaanxi Province built a "One-Map" system for basic information resources to strengthen integration and management of the basic information resources of government affairs in fields such as demography, legal persons, geographic space and macro-economy, and provide support services for government departments at all levels and various industries. Ningxia strengthened its construction of the big data service platform for government affairs. Specifically, a unified website for big data analysis and people's livelihood services was established based on the data sharing and exchange system for government affairs data and the big data analysis system, which provided more convenient and intelligent means for the government to manage decisions and serve the masses.

4.3 Solid Progress in "Internet + Government Services"

The "Internet + government services" construction is a key measure to deepen the reform of "streamlining administration & delegating power, strengthening & loosening regulation, and optimizing service", and is of great significance to speed up the transformation of government functions, improve the government administrative efficiency and enhance people's well-being. In recent years, China has actively advanced "Internet + government services" and made remarkable achievements in building the National Integrated Online Government Service Platform, innovating service patterns, expanding service channels and diversifying service principals.

---

[2] As of the end of 2018.

4.3.1   The National Integrated Online Government Service Platform Starting Online Operation

The National Integrated Online Government Service Platform is an important carrier to achieve the national unified standards for and whole-process online handling of government service items, and promote cross-regional, inter-departmental and cross-level data sharing and business collaboration in respect of government services. In 2018, China issued *Regulations on Issuing the Implementation Plan for Further Deepening "Internet + Government Services" and Promoting the Reform of One-Stop Online Government Services by One Department at One Time, Guiding Opinions on Accelerating the Advancement of the Construction of National Integrated Online Government Service Platform* and other policy documents to actively promote the establishment of the National Integrated Online Government Service Platform.

In January 2018, the project of building the National Integrated Online Government Service Platform officially started, which mainly included the construction of the big database for government services, service and work portals, public support systems, key applications and management standard systems. In May 2019, the National Integrated Online Government Service Platform (see Fig. 4.1) was put into trial operation. The platform acts as a public entrance, public channel and public pillar and provides "seven unified" services, namely, unified identity authentication, unified certification service, unified matter handling, unified complaints and suggestions, unified rating, unified user service and unified search service, for the government service platforms of all regions and all departments across the country. In doing so, its four major functions are realized, namely, supporting one-stop services, gathering data and information, promoting data exchange and sharing, and strengthening dynamic supervision. In addition, outstanding issues in cross-regional, inter-departmental and cross-level government services, such as difficulty in information sharing and business collaboration and lack of basic support, have also been solved. As of July 2019, 1,142 government service items and 308 convenience service applications of 46 State Council departments, and more than 1.93 million government service items and 532 convenience service applications of 31 provincial administrative regions and Xinjiang Production and Construction Corps had been included in the National Integrated Online Government Service Platform.[3] This has effectively propelled the regulated, standardized and intensive building and interconnection of the government service platforms of all regions and all departments, and provided effective platform support for accelerating the building of a comprehensive system for government services across the country.

---

[3]Data Source: China's National Integrated Online Government Service Platform, http://www.gjz wfw.gov.cn/.

Fig. 4.1 China's national integrated online government service platform

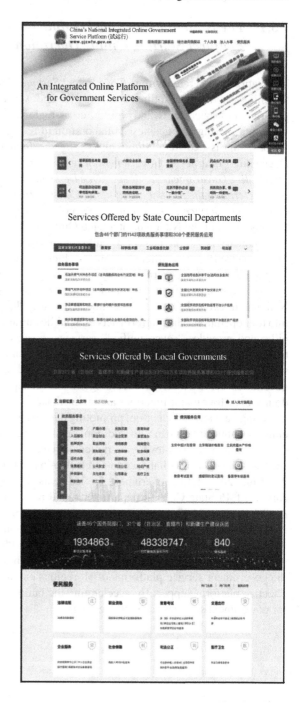

4.3.2 Continuous Innovation in Online Government Service Patterns

Enterprises and the public strongly express that arranging matters is difficult, slow and complex. In order to resolve the problems like this, all regions and all departments have actively explored "Internet + government services" by sticking to the problem-oriented and demand-oriented principle and innovating government service patterns. As a result, a great many of advanced models have been produced. Zhejiang's "one visit at most", Jiangsu's "online approval", Guizhou's "five comprehensive services" and Shenzhen's "swift approval in seconds", among others, have aroused great response throughout the country.

Zhejiang was the first in the country to launch the "one visit at most" pilot program. By relying on the Zhejiang Government Affairs Service Network (see Fig. 4.2), it broke the data barriers and strengthened data sharing among the information systems of different departments, so that the masses needed not to visit more than one departments for many times to handle their businesses. As of May 2019, Zhejiang had announced 1,411 main items and 3,443 sub-items of "one visit at most" at provincial, municipal and county levels. It has continued its efforts to promote online handling of all the people's livelihood service items and enterprise service items. It is expected that by the end of 2019, more than 60% of government service items would be handled online and more than 70% of people's livelihood items would be handled by presenting only the ID card.[4] Meanwhile, Zhejiang has improved its matter handling and consultation service system, updated the supporting knowledge base, strengthened multi-channel publicity and provided convenient and efficient consulting services by relying on the administrative service centers & government service websites at all levels and the "12345" government hotline intelligent customer service platform. Consequently, it has significantly improved the convenience, accessibility and accuracy of matter handling and consultation services, and greatly relieved the issue of "many visits for one matter".

Guizhou has focused on the regulated and intensive building of government services and proposed to build a provincial integrated online government service platform by developing the model of "five comprehensive" services, namely, full-coverage, all-connected, all-round, all-weather and all-process services, so as to realize the goal of "handling matters across the province within one system". In the first half of 2019, more than 400,000 service items of more than 140,000 departments at the five levels of province, city, county, township and village were disclosed and handled intensively on the platform, and all relevant data was migrated to the "Guizhou-Cloud" platform. In doing so, all the government service items within the provinces were handled on one website and the departments at city, county and province levels needed no longer to build their own approval systems. A highly integrated and intensive model for the construction of government services has been

---

[4]Xinhuanet, As the Lists of "One Visit at Most" Items Have Been Released by Many Regions, Do the Public Need to Repeatedly Visit Government Departments? http://www.xinhuanet.com/pol itics/2019-05/15/c_1124497454.htm.

Fig. 4.2 Zhejiang government affairs service network

Fig. 4.3 Guizhou government service platform

established, which has greatly contributed to the transformation of government functions, the integration of government data and the improvement of administrative efficiency,[5] as shown in Fig. 4.3.

Shenzhen has carried out the reform of "swift approval in seconds". It has explored ways to achieve automatic online examination and approval on the basis of data sharing, item standardization and uniform approval rules. That is to say, the applicants first submit relevant information on Internet, then the system will compare and check them in real time through data sharing according to specified rules, automatically make the approval decision, and notify the applicants of the result in a timely manner. The reform of "swift approval in seconds" has achieved remarkable results in promoting the transformation of government functions, optimizing the business environment, improving the administrative efficiency, reducing the cost of businesses handling, preventing rent-seeking behaviors and enhancing the governance level. As of June 2019, Shenzhen had realized "swift approval in seconds" for 52 service items in talent introduction, application for old age allowance, application for online car-hailing/taxi driver's license and social investment project filing, which resolved the bottlenecks and sufferings in handling matters for the masses and put an end to human disturbance. Took the introduction and settlement of university graduates as an example. After the reform of "swift approval in seconds", university graduates need only to submit their information on the Internet, and the system will automatically examine and approve it and report the result immediately. Finally, they will

[5]Data Source: http://www.gzegn.gov.cn/.

complete their settlement procedure simply by appearing at the household register window. From June 2018 to March 2019, more than 64,000 non-native university graduates in Shenzhen obtained Shenzhen's *hukou* (permanent residence) through the "swift approval in seconds".

4.3.3 Online Government Service Channels Becoming More Diversified

To actively promote the handling of government service matters with wide coverage and high-frequency applications on mobile terminals and thus realize their mobile handling by clicking with fingertips is an important content of improving government services for the convenience and benefit of people. With the rapid development of mobile Internet, new technological applications such as government affairs apps and WeChat-based government service applets that are developed under the leadership of the government, have increasingly become a new channel for the public and enterprises to obtain government information and handle service matters. As of July 1, 2019, a total of 31 provincial government service mobile terminals had been built by 31 provinces (autonomous regions, the municipalities directly under the Central Government) and Xinjiang Production and Construction Corps. As mobile applications have been constantly emerging, such as Guandong's "Yueshengshi", Zhejiang's "Zheliban", Shanghai's "Suishenban", Chongqing's "Yukuaiban", Anhui's "Wanshitong", Beijing's "Beijingtong", Fujian's "Minzhengtong" and Shanxi's "Sanjintong", mobile government services are entering a period of accelerated development.

Guangdong Province has actively built the "Yueshengshi" platform on WeChat as a unified channel of offering e-government services, as shown in Fig. 4.4. The Yueshengshi platform aims to build an online "Whole of Government" (WOG) based on the "separation of matters accepting and handling and information communication". The back-end system of the platform connects with more than 100 business systems of government sectors through real-name, trusted, face-scanning identity authentication of the Ministry of Public Security, and provides one-stop services for consultation and handling of 541 high-frequency service matters. It has effectively resolved the issue of separate operation of different departments and broken the inefficient model of separate development, operation and maintenance of government mobile applications, thus realizing inter-departmental, cross-level and unified acceptance of government service matters. At the front end of the platform, the public can directly visit Yueshengshi's matter handling interface on WeChat, which has effectively solved issues such as problems in finding and popularizing government service mobile apps as well as their low usage. In just two months after its launch, the public security section of the "Yueshengshi" platform has provided 8.59 million services, 28 times more than that of the original app.[6]

---

[6]Data Source: https://www.digitalgd.com.cn/product/439/.

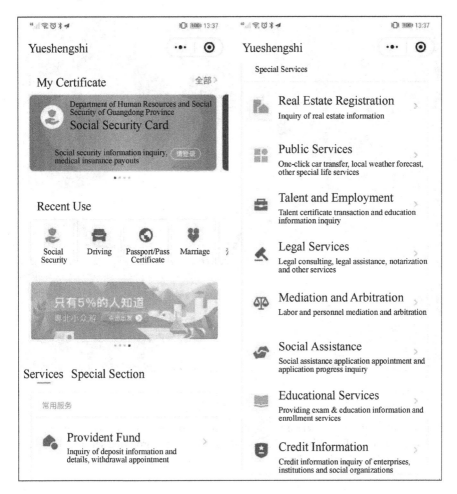

Fig. 4.4 Yueshengshi platform

Zhejiang Province has vigorously promoted mobile Internet government services and developed the "Zheliban App", the mobile client for government services across the province, to realize the handling of government service matters by clicking with fingertips rather than personally visiting government departments, as shown in Fig. 4.5. Matters such as hospital registration, traffic offense handling, applying for property ownership certificates, and even subway ticket purchase, and hunting for parking spots and public toilets can all be handled conveniently on the "Zheliban App". In the first half of 2019, all the administrative service centers in the province opened the function of mobile business handling on the App, with a total of 15,853 service matters included on the app. Zhejiang also took the lead in promoting the reform of electronic collection of government non-tax revenues and built a unified public payment platform for government services that covered hundreds of payment

Fig. 4.5 Zheliban app

matters, including traffic violation fines, registration fees for license examinations and social security fees. Since its launch, the platform has provided services for 87.88 million people and collected a total of ¥ 114.174 billion.[7]

4.3.4 A Greater Variety of Entities Participating in E-government Construction

To support and guide social forces to actively participate in e-government construction is of great significance to improve the level of e-government construction and enhance the government's public service capability. In recent years, the cooperation between government and firms in e-government construction has become more extensive. Through active engagement in e-government construction, traditional information technology enterprises such as software/hardware manufacturers and operators as well as Internet enterprises are becoming important providers of e-government

---

[7]Data Source: http://www.zjzwfw.gov.cn/.

services who have brought new ideas to e-government construction and pushed e-government into a new stage of development. At present, business engagement mainly focuses on the following two aspects.

(1) In terms of project construction, both traditional information technology enterprises and Internet enterprises such as Alibaba and Tencent are actively participating in e-government construction projects, such as the "Digital Guangdong", the "Action Plan of Beijing Municipality" for the Development of Big Data, the "Guizhou-Cloud" platform, the "Golden Gate Phase II" Big Data Cloud Project, Hainan's "Urban Brain", and Changsha's "Urban Super Brain". The active participation of various enterprises provides strong technological support and capacity assurance for e-government construction.

(2) In terms of service channel, the government and platform enterprises have deepened their cooperation and fully utilized Alipay, WeChat and other platforms to provide the public with government services. For example, the "City Services" section was set up on the platforms, and WeChat public accounts and applets for government services were launched to provide services such as social security information inquiry, tax declaration and peccancy disposal for users. This model of providing services through third-party channels is being recognized by an increasing number of people. According to the 44th "Statistical Report on China Internet Development" released by the China Internet Network Information Center (CNNIC), as of June 2019, WeChat-based urban services were available in 31 provinces (autonomous regions, municipalities directly under the Central Government), and the number of people who received governmental services via WeChat was 620 million. Figure 4.6 shows the total number of the users of WeChat-based urban services from December 2017 to June 2019.

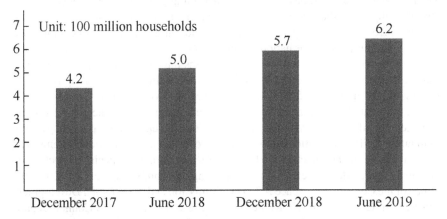

Fig. 4.6 Total number of the users of WeChat-based Urban services from December 2017 to June 2019

4.4 Efforts Stepped up in the Integration and Sharing of Government Information Systems and the Open Access of Public Information Resources

In order to help resolve the issues in government informatization construction, such as "fragmented administration, sectoral fragmentation, disorganized management and information islands", since 2017, China has deepened the integration and sharing of government information systems and the open access of public information resources, and remarked achievements have been obtained as a result.

4.4.1 Steady Progress in the Integration and Sharing of Government Information Systems

In 2017, *Notice of the General Office of the State Council on Issuing the Implementation Plan for the Integration and Sharing of Government Information Systems* was issued, marking the beginning of the work of integration and sharing of government systems in China. After that, the National Development and Reform Commission (NDRC) and other relevant departments have issued a series of documents in succession, including *Notice on Issuing the Work Plan for Accelerating the Implementation of the Implementation Plan for the Integration and Sharing of Government Information Systems, Notice on Carrying out the Pilot Projects of Integration and Sharing of Government Information Systems,* and *Notice on Further Accelerating the Integration and Sharing of Government Information Systems.* In order to continuously advance relevant work, it has convened four leading group meetings for promoting the integration and sharing of government information services and more than 20 working meetings of the organization and promotion group for the integration and sharing of government information systems.

A number of government information systems have undergone rectification and standardization. As of June 2019, China had systematically combed through the basic information of more than 5,000 information systems of the State Council departments, eliminated more than 2,900 information islands, and connected 42 vertical management information systems of the State Council departments, providing a solid foundation for the integration and sharing of government information systems. The General Office of the State Council took the lead to conduct special inspection on the State Council departments' work of cleaning up "zombie systems". In 2018, the Office of the Central Cyberspace Affairs Commission and the National Development and Reform Commission (NDRC) evaluated the work of 61 State Council departments and units on the integration and sharing of government information systems. Specifically, they deeply investigated the progress in the integration and sharing work, the integration of information systems, the sharing of information resources, business collaboration and the system of government service platforms, and comprehensively evaluated the effectiveness of the integration and sharing of government information systems. They also analyzed the existing problems and shortcomings to provide reference for the further integration and sharing work.

The national system for government data sharing and exchange has been gradually established and improved. China has actively advanced the integration and sharing of government information systems through building a "broad channel" for information sharing, a "general catalog" for government data resources and a "overall hub" for data sharing and exchange. The national system for government data sharing and exchange focuses on building a comprehensive network platform featuring "one network connection, one cloud bearing, one platform sharing, one portal opening, one directory guidance, and one certificate recognition" and promoting "network communication, data communication and business communication", which are collectively called "Six One Promoting Three Communications". At present, China has initially built a national government public network that has the largest network coverage, connects to the largest number of government departments, provides the most comprehensive business services, and contains the richest government information resources in China. It enables the sharing and access of basic data about the population, legal persons, spatial geography as well as data in key areas, and provides a favorable platform environment for the integration and sharing of national government information systems, the construction of the "Internet + government services" system, and the safe and controlled aggregation, development and utilization of government big data.

Besides, all regions in China have made active explorations to promote the sharing and integration of government information resources and break information islands. For example, the "Digital Fujian" project pooled together more than 3,200 pieces of data and 2.1 billion records from more than 100 provincial and municipal units. In Ningxia, the government big data service platform connected with 818 shared directories and 130 million pieces of shared data at province, city and county levels, and the system for government data resources sharing was basically established.

4.4.2 Accelerating the Opening of Public Information Resources

The opening of public information resources is an important step to make information services more beneficial for people, promote the scale, innovative application of information resources and strengthen digital economy. In 2017, the Office of the Central Cyberspace Affairs Commission, the National Development and Reform Commission (NDRC) and the Ministry of Industry and Information Technology (MIIT) jointly issued *Work Plan for the Pilot Program of the Opening of Public Information Resources*. The Plan proposes to address the main difficulties in the current work of opening public information resources, such as the lack of a unified platform and insufficient data application, management standardization and safety assurance by launching pilot projects in five areas including Beijing and Shanghai. The projects focus on six aspects of work, namely, building a unified open platform, clarifying the opening scope, improving data quality, promoting data utilization, establishing and improving relevant systems and norms, and tightening security.

Since the launch of the pilot work, the public information resources have quickened the pace of open exchange. As of June 2019, 82 provincial, sub-provincial and prefecture-level governments had established data opening platforms, an increase of 36 local platforms over 2018. Specifically, data opening platforms were set up in 41.93% of the provincial administrative regions, 66.67% of the sub-provincial cities and 18.55% of the prefecture-level cities in China. The government data opening platform has gradually become an essential element in the construction of local digital government. The total quantity of open data sets across the country has increased rapidly from 8,398 in 2017 to 62,801 in 2019, nearly a sevenfold increase. The volume of open data sets has witnessed explosive growth to nearly 20 times from June 2018. About 30% of the platforms have more than 1,000 open data sets.[8]

All regions and all departments have made active explorations in the opening of public information resources. For example, Shanghai has actively carried out the pilot project, developed more than 2,000 open directories and clarified the content, forms, updating frequency and attributes of data opening. The open data has been disclosed to the public on the Shanghai Government Data Service Network. The scope of data opening keeps expanding to cover 12 key fields (e.g., economic construction, resources and environment, and education and technology) and 11 use cases (e.g., school education and lifelong education, government affairs handling and the disabled services).[9] The State Forestry Administration has actively explored new ways of opening the government data on forestry. The Forestry Data Opening and Sharing Platform of China (data.forestry.gov.cn) provides data retrieval, data statistics and data analysis functions through sharing of data, information and resources, thus making in-depth mining and customized collection of data possible. As a result, a database containing 3 categories, 9 sections and 17 subcategories of key data has been established, which has satisfied the demands of multi-information collection, multi-type services and multi-channel access. The public can query against the database by data type, topic and form. It has become an authoritative thematic data platform of the forestry industry.

The public and enterprises have fully utilized the government's open data to innovate and expand public services. One example is the mobile app "Visit Museums" ("*guang guang bo wu guan*") developed based on the open data of Beijing's government data resource website (www.bjdata.gov.cn), which provides audio guide and indoor positioning of many museums in Beijing. Another example is the mobile app "Parking" ("*ting che*") developed based on the open data of Guiyang's government

---

[8]Data Source: Digital and Mobile Governance Lab (DMG) of Fudan University, *China Open Data Index in 2019*.

[9]People.cn, Remarkable Results Are Achieved in Shanghai's Pilot Project for Promoting Opening Public Information Resources, http://sh.people.com.cn/n2/2019/0219/c134768-32657680.html.

data opening platform (www.data.guiyang.gov.cn). The app enables the sharing of parking information resources and creates a new parking management model named "Parking Supermarket".

4.4.3 Remarkable Achievements in the Integration and Sharing of Information Resources

In order to fully tap data potential and unleash data value, all regions and all departments in China have actively explored the mechanism for integration, sharing, development and utilization of public information resources. Some provinces and cities have made breakthroughs in other fields such as poverty alleviation, social assistance, business credit investigation, and operational & post-operational oversight.

In order to address the issues of scattered statistical information on poverty alleviation, data distortion and inefficient data mining and utilization, Guizhou, Guangdong and other regions have actively built big data support platforms and broke down information islands among poverty-alleviation governments by establishing interconnected poverty-alleviation information systems, thus realizing "precise management" of the poverty alleviation work. Guizhou has established connectivity among 13 departments, which covers the department of public security, the department of industry and commerce and the department of transportation and makes one-click inquiry of the data of more than 7 million registered poor households in Guizhou Province possible. Guangdong's unified government information resources sharing system pools the basic data of 664,000 poverty-stricken families and 1.731 million poor people, which is combined with the data of online services and government affairs in industry sectors. The system incorporates the data from the departments of civil affairs, education, and human resources and social security, which is accessible at the province, city, county, township, village and household levels. There is currently a total of 330 million pieces of shared information in the system.

Facing the issues of incomplete data on urban and rural subsistence allowances, time-consuming and labor-consuming data verification, and difficulty in preventing "relation assurance" and "favor assurance", Guangxi, Qinghai and other regions have strengthened inter-departmental cooperation on social assistance and built integrated databases by integrating the data of social security, civil affairs, housing construction, public security and other departments. Meanwhile, they have also worked resolutely to "kick" those living in abundance who arouse strong public resentment out of the coverage of subsistence allowances. For example, Guangxi has integrated the information of 15 government departments, including the civil affairs bureau, court, public security bureau and human resources and social security bureau, and built the provincial big data platform for online verification of subsistence allowance data at district, city and county levels. Upon entering into the data exchange agreement, relevant departments may share data in both directions and in real time, with data exchange completed within two minutes upon request. Qinghai has greatly improved the accuracy of relief objects identification through data exchange between the platform for checking the financial conditions of households and 15 government departments, including the provident fund department, census register department, and individual income tax department.

To solve the difficulties in the integration and sharing of credit information, such as bank-enterprise information asymmetry, blockage of the "information sharing" mechanism, insufficient application of business credit information, Jiangsu, Fujian and other regions have vigorously built credit information service platforms and established the systems for integrated sharing and diversified data application. Suzhou, a city of Jiangsu Province, has built a credit information platform to effectively gather the government's business-related information and enhance the credit of enterprises. As of the end of 2018, it had collected 66 million pieces of credit information from government departments and public institutions as required, 160,000 small and medium-sized enterprises in the city had their credit information registered on the platform, 89 financial institutions had established cooperation with local credit platforms, and the platform had received a total of 266,000 queries. Fuzhou, a city of Fujian Province, has also established a big data credit platform, which has incorporated a total of 1.35 billion data records of 2,800 categories from 55 provincial departments and 50 prefectural and municipal units. The platform now connects with 58 micro-loan and financing guarantee companies, with the credit codes of 863,000 institutions collected and 247,000 credit report queries handled.

In response to issues such as the lag in the operational and post-operational oversight emerging after the reform of the business system, and the regionalization, decentralization and fragmentation of credit information, Zhejiang and other regions have promoted the interflow of regulatory information among different departments, fully uncovered the big data value of regulatory information, and improved the refined and scientific level of operational and post-operational oversight. Hangzhou, capital city of Zhejiang Province, has established a platform for joint supervision of enterprise credit. As of the end of 2018, it had collected 61.13 million pieces information on dishonest persons in the daily supervising work of 39 departments, which had been matched with corresponding enterprises. In addition, they were marked in different colors to display the regulatory status. In specific, 9,375 enterprises were on the "Credit Prompting" list, 14,000 on the "Credit Warning" list, 60,800 on the "Abnormal Operation" list, and 3,366 on the "Credit Restriction" list. Besides, 2,066 dishonest persons subject to enforcement had been intercepted.

4.5 Technological Innovation Playing a More Important Role in Supporting E-government Development

With the continuous development of big data, AI, IoT, blockchain and other information technologies, they are showing evident advantages in perceiving social trends, unblocking communication channels and assisting scientific decision-making. All regions and all departments in China have actively promoted the application of the new-generation information technology into the key fields of government affairs, and continued to enhance the innovation in e-government service and management models

4.5.1 IoT Technology Helping Improve the Government's Capacity of Digital Perception

IoT technology enables ubiquitous environmental perception and situational monitoring with Internet of Everything (IoE), and is now widely applied into government supervision, urban governance and other fields. For example, Inner Mongolia has established an ecological environmental data resource center, which combines IoT technology and traditional methods to develop and integrate 110 items of data in 9 categories on the automatic monitoring of pollution sources and environmental quality, accounting for 90% of the environmental protection data. It has provided great convenience for carrying out the environmental protection work. Guizhou has established a big data and IoT traceability cloud platform to record the production information of 495 kiwifruit growers in Xiuwen County, including the basic information of the orchards and the records of fertilization, pesticide use, fruit growth and fruit picking. Besides, in order to ensure food safety, it provides traceability management for the whole production process through QR code scanning. The platform currently provides safety pre-warning, source traceability, movements tracking, information inquiry, responsibility identification and product recall for 236 orchards in the county, which cover an area of 51,270 μ or 31.9% of the total kiwifruit planting area in Xiuwen County.[10] Besides, in the transportation field, a great many of cities have installed IoT and monitoring devices to obtain real-time traffic information so that timely and correct traffic control measures could be deployed.

4.5.2 Big Data Making the Government's Decision-Making More Scientific

Big data technology analyzes, mines and extracts valuable information from massive data, thus creating conditions for the government's scientific, efficient, accurate and rapid decision-making, as well as services and supervision. On the one hand, it enables data recording of the government's whole work process, and further supports deep analysis, process backtracking, post supervision, service optimization and other

---

[10]Guizhou Big Data Development Administration Website, Two Agricultural Big Data Application Projects of Guizhou Province Were Selected into the National Top Ten Application Cases of "Big Data + Poverty Alleviation", http://dsj.guizhou.gov.cn/xwzx/gzdt/201905/t20190527_343 6043.html.

related work. On the other hand, massive data is rapidly collected and sent to government regulators by means of information technology, and provides information support for the timely detection and disposal of relevant problems.

4.5.2.1 Big Data Has Effectively Improved the Scientific Level of Government's Decision-Making

For example, in terms of city operation and management, big data technology can be used to visually present the city operation status, allowing city administrators and decision makers to follow the progress of relevant events and find out the potential problems in city operation. This provides scientific basis for leadership decision-making and policy formulation. Many regions, such as Beijing, Guangdong and Guizhou, have actively used big data resources, technologies and platforms to explore the value of data resources, transform the government operation model and build a new-type government that makes judgment and decisions based on data, trying to make data the "microscope, photoscope and telescope" for government governance.

4.5.2.2 Big Data Has Greatly Improved the Government's Livelihood Services

Big data can be used to explore the value of government data resources, making the provision of accurate and personalized public services possible. Guangzhou, Shenzhen and other cities have identified users' potential demands by analyzing the data on their personal attributes, previous services, etc., and thus provided accurate and personalized services for them on the government websites to better satisfy their demands of livelihood services.

4.5.2.3 Big Data Has Driven Constant Innovation and Upgrading of the "Internet + Supervision" Model

The application of technologies based on big data analysis and image recognition offers strong technological support for government supervision, public security and other work, which has increased the efficiency of government affairs and social security. For example, Guizhou Province has launched the "Data Cage" project to give play to the data's platformization, correlation and cohesion role in application-oriented supervision, inspection and technological anti-corruption. It has become an important means to improve the government's governance capacity and innovate the pattern of government supervision. Guiyang Traffic Administration Bureau, as a pilot region of the "Data Cage" project, has integrated 22 business subsystems and built one integration platform, one mobile App, one personal credit record, and more than 20 models for business risk pre-warning according to the "3+N" model after more than three years of efforts. The supervision platform automatically extracts data on a daily basis to predict, prevent and warn against the risks emerging in the government's exercise of power. Thanks to real-time entry of information, the whole process of law enforcement against drunk driving is monitored, which has effectively addressed the loophole of depending on human relations. Meanwhile, through the "Data Cage" system, the platform can automatically judge whether the police are working, which has improved the accuracy of supervision and greatly reduced the manpower and material costs of supervision. This has effectively changed the traditional administration model of "human management, supervision and execution" and transformed "human supervision into data supervision" and "passive supervision into active supervision".

4.5.3 AI Enhancing the Intelligence Level of Public Services

AI technologies are transforming the government's existing management and service model. In recent years, AI technologies such as pattern recognition and semantic analysis have achieved constant breakthroughs, the application of AI service platforms and interfaces has become increasingly extensive, and intelligent customer services and other models have rapidly spread in the business domain. In the e-government field, AI is widely used to assist collecting information and screening intelligently, accepting fuzzy tasks and performing pattern recognition, which plays an important role in improving the government's work efficiency, service capability, decision-making level and the experience of interaction between government and the public.

4.5.3.1 Matters Are Automatically Handled at Intelligent Terminals

As AI technology and terminal device continue to develop and mature, "7×24" self-service business handling has become possible for government services. In November 2018, a 24-hour self-service hall at the Xuhui District Administrative Service Center in Shanghai was officially opened to the public. In the hall, robots are available to provide help for the public in self-service inquiry, self-service processing, self-service printing, self-service pick-up, self-service logistics and other services, making it a "7×24" unmanned supermarket of government services. Another example is the "7×24" unmanned station for government services in Leshan City of Sichuan Province. It has installed AI intelligent terminals to identify the operation intention of the public, conduct real-name authentication for them through voice feedback, and collect and transmit their materials. The station is maintained on duty 24 h per day, 365 days a year.

4.5.3.2   Intelligent Question Answering Systems Respond to Users Rapidly

In recent years, the intelligent question answering technology has been applied into e-government work. In 2017, "Jingjing", an intelligent question answering robot, was launched on Beijing Municipal Government's website "The Window of the Capital" (see Fig. 4.7), which answered questions related to the political dynamics, policy documents and their interpretation, matters processing and processing guidelines on the website. It currently possesses more than 110,000 pieces of information. The robot comprehends users' natural language questions and intentions, answers their questions rapidly and correctly through friendly question answering process, and offers recommendations accordingly. It also collects user satisfaction and continues to optimize and improve its knowledge base through deep learning, which has greatly improved the search accuracy and reduced users' search time. Since its launch, "Jingjing" has been used by more than 160,000 users and answered more than 220,000 questions with an accuracy rate of 98%, making government services much more beneficial to the public.

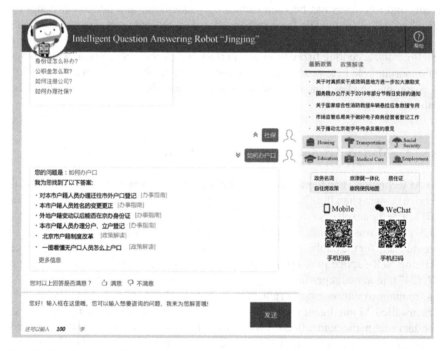

Fig. 4.7 Jingjing, an intelligent question answer robot on beijing municipal government's website

4.5.4 Blockchain Technology Helping Creating an Environment of Sincerity and Honesty

Due to its decentralization, transparency and openness features, blockchain technology can help resolve the problems in identity recognition and information integrity, reliability and security of government services. It has extensive application prospects in fields such as user authentication and security assurance. For example, it can be utilized to confirm the identity of the public and enterprises, and thus reduce the materials to be submitted and simplify the certification process. Besides, it can also be applied into tax, house management, market supervision, healthcare and other fields to ensure the authenticity, effectiveness and traceability of the information related to funds, property rights, credit, health, etc.

In order to address the issues of not sharing data between departments for fear of data leakage, insecurity and lost of control, Shaanxi Province has explored a new model of government data application featuring clear responsibilities and rights and high security and reliability according to the technological characteristics of blockchain and intelligent contract. In the new model, the data used always remains in the system and the information usage can always be traced. Other efforts include strengthening regulatory capabilities, establishing consensus standards, promoting data fusion and exploring data value. In Xianyang City alone, blockchain technology has been used to retrieve data on the poverty conditions of 290,000 people and the health conditions of 4.92 million people, which is reliable enough to be applied. According to the data, the city has identified 123,000 people living in poverty, over-fulfilled the poverty alleviation task by lifting an extra 7,000 people out of poverty, eliminated 320 disqualified people out of and added 1,512 people onto the list of poverty alleviation targets. Meanwhile, it has worked to resolve the difficulties in the medical reform, saving ¥ 67.2 million of medical insurance funds in total and reduced ¥ 121 million of medical expenses.

Nanjing has utilized blockchain technology to handle the matters of real estate transaction and real property registration, and built a unified database for data recording, which has promoted safe and efficient information maintenance. Inter-departmental information sharing is completed on an electronic certificate platform based on blockchain technology, which provides traceability for relevant data through its whole life cycle. The data can only be read and written, and may not be modified or deleted. Besides, a unique consensus mechanism of multi-party signature endorsement and multi-party query verification has been established and comprehensively applied into house-purchase certification, real estate transaction, real property registration and other business links. On the blockchain-based electronic certificate platform, the Real Estate Bureau and the Land and Resources Bureau jointly execute the smart contract based on common business requirements and maintain the registration, pledging and transaction records of each piece of real estate data. It has effectively solved the problems in real property transaction and ensured the security and efficiency of data and information sharing among transaction parties.[11]

4.6 Further Consolidation of the Basic Support and Assurance System

In recent years, with the continuous development of information technology in China, its capacity of providing basic support for e-government development and the intensive level of infrastructure and application have been greatly improved. Besides, China has continuously strengthened its capacity to protect cybersecurity and expanded the talent teams, which lays a solid foundation for healthy and safe e-government development.

4.6.1 Significant Rise in the Intensive Level of Infrastructure and Application

Government information infrastructure is the basis of e-government application. Chinese government has prioritized the task of infrastructure construction, continued to improve government network infrastructure, and strengthened its capacity of supporting e-government development.

China's government network infrastructure has been continuously improved. A national public government network featuring the largest network coverage, connecting with the largest number of government departments, and providing the most comprehensive business services has been built. The network has basically achieved full coverage at the four levels of central government departments, provinces, municipalities, districts and counties with sufficient reach and depth. It provides a favorable environment of platform facilities for the integration and sharing of national government information systems, the construction of the "Internet + government services" system, and the safe and controlled aggregation, development and utilization of government big data. As of December 2018, more than 40 national ministries had provided services on the government network, including public services, intra-government services and basic services.

Intensive construction has been a focus of e-government development in recent years. To a certain extent, it has solved the issues of disorderly investment of e-government funds, repeated system construction and low resource utilization rate. On the one hand, the intensive construction of government cloud keeps accelerating. At the level of central ministries, the Ministry of Civil Affairs has implemented dual

---

[11] The Chinese Central Government's official website, [Service Optimization] Nanjing Pioneered the Model of Integrated Business Handling for Real Estate Transaction and Real Property Registration, http://www.gov.cn/zhengce/2017-10/27/content_5234861.htm.

active storage and cluster integration and upgrading of government cloud platforms, which has greatly improved their business support capacity and ensured their availability and data security. By taking advantage of the national intensive platform for information release, the Ministry of Commerce has built institutional sub-stations for ministry leaders, ministry departments, special commissioner's offices stationed at different localities, economic and commercial institutions based abroad, public institutions directly under this Ministry, chambers of commerce, academies, associations and other units. At the local level, more than 20 provincial administrative regions, including Beijing, Shaanxi and Ningxia, have engaged in government cloud construction, and concentrated on various software and hardware resources such as machine rooms, storage device and OA system on the cloud. For example, in 2018, 8 cloud service providers moved to the Beijing municipal government cloud to provide cloud services for nearly 800 service systems of more than 60 commissions, offices and bureaus of the municipal government, which effectively promoted data sharing and business collaboration among all departments and all regions in China. Shaanxi incorporated 1,233 business application systems and hundreds of databases of 95 provincial, 11 prefecture-level and 107 county-level departments on the government cloud platform. Intensive construction has helped save 55% of the construction funds and 60% of the operation and maintenance costs. As of July 2019, the Ningxia e-government public cloud platform carried 899 application systems of 289 government departments. The goal of provincial concentration of government information systems had been initially realized.

On the other hand, the intensive construction of government websites has achieved remarkable results. In 2018, the General Office of the State Council issued *Plan for the Pilot Work of Trial Intensification of Government Websites*, stipulating that the pilot work of trial intensification of government websites would be carried out in 10 provincial administrative regions including Beijing, Jilin and Shandong and Lhasa city of Tibet Autonomous Region. The pilot project has promoted better integration of resources, secure platform integration, mutual recognition and sharing of data, overall and standardized management, and the improvement of service convenience and efficiency of various government websites at all levels. For example, the Guizhou provincial government website platform has provided support for the intensive construction of the government websites across the province, including their overall migration, hierarchical integration and standardized construction. As of the end of 2018, the Qingdao government website integrated millions of pieces of information regarding institutional functions, official documents, law enforcement and affairs handling of 73 municipal departments. Besides, a "one-stop government channel" for handling matters such as public consultation, help seeking, criticism, suggestions and complaints has been established, with the number of annual accepting cases exceeding 100,000. More than 4,000 network politics campaigns have been conducted and participated in by more than 1.9 million Internet users who had about 140,000 questions answered. In addition, thousands of "golden ideas" have been incorporated into the municipal government's decision-making or departmental work plans.[12]

4.6.2 Continuous Improvement in the Capacity to Protect Cybersecurity

After years of development, e-government has penetrated into every aspect of economic and social development in China. With a huge amount of data and information on national political security, economic security, national defense security and social stability, the e-government information system has a bearing on the healthy e-government development and constitutes an important component of the national security system. Effective steps must be taken to establish and improve the e-government information security assurance system, comprehensively enhance the capacity of protecting information security, focus on protecting the security of basic information networks and important information systems, and create a safe and healthy network environment. These measures would promote healthy and safe e-government development and safeguard public interests and national security.

In recent years, China has continuously enhanced its capacity of guaranteeing e-government security. Relevant departments formulated and issued *Measures for Data Security Management (Draft for Comments)* for public comment, improved the network and information security infrastructure, and conducted monitoring and reporting of the cybersecurity of Party and government organs' websites. In the meantime, efforts have been stepped up in protecting personal privacy in the use of government information resources and guaranteeing the safety of government information resources during their collection, sharing and use.

4.6.3 Intensified Efforts in Talent Team Construction

4.6.3.1 The E-government Expert Consultation Mechanism Is Initially Established

In 2018, the Office of the Central Cyberspace Affairs Commission and other relevant departments set up the National E-government Expert Committee, which was responsible for studying major issues in e-government construction and management across the country, guiding all localities to carry out comprehensive pilot projects for e-government, identifying the trend of e-government development, and providing advice for the formulation of the national strategic plan for e-government development and the construction of major projects. All regions have actively pushed forward the establishment of the e-government expert consultation mechanism. As of May

---

[12]Data Source: He Yiting, *Annual Report on China's E-Government Development (2018–2019)*.

2019, 10 provincial administrative regions, including Beijing, Fujian and Xinjiang, had established their own e-government expert consultation institutions to provide a range of professional services, including e-government planning, project review, policy consultation and formulation of technological standards and regulations, and to support scientific decision-making.

4.6.3.2 Trainings and Exchanges Regarding E-government Are Carried Out on an Extensive Scale

E-government has now become a part of civil servant training. *National Plan for Cadre Education and Training (2018–2022)* incorporated the training of new knowledge and skills related to Internet, big data, cloud computing and AI into the national cadre training plan. Since 2017, the Office of the Central Cyberspace Affairs Commission has organized national e-government training courses every year to study and implement General Secretary Xi Jinping's important thought on building China's strength in cyberspace and provide centralized training on accelerating the national e-government construction. In 2018, the General Office of the State Council convened the Mobilization and Deployment Meeting for the Construction of the National Integrated Online Government Service Platform and Centralized Training Meeting, on which it proposed to accelerate the construction of the National Integrated Online Government Service Platform, and further promote the construction of "Internet + government services". Forums on e-government, "Internet + government services" and other related topics have also been held at important conferences such as the Digital China Summit and the World Internet Conference, to display the e-government development cases, exchange the experience of e-government construction, and explore new models of e-government development. In addition, e-government courses and specialties have been set up in many Chinese colleges and universities. As of the end of 2018, more than 50 colleges and universities established the e-government major or research direction, and more than 100 colleges and universities opened e-government courses.

4.6.3.3 Talent Team Construction Is Continuously Intensified

In terms of talent team construction, all localities in China have broadened the channels for personnel selection and appointment, established and improved the talent incentive mechanism, and built professional teams with proficiency in government services and digital strategic thinking. Take Hainan Province for an example, the Big Data Management Department of Hainan Province has implemented the post management system, according to which the department may open jobs for top specially-employed positions such as big data architect according to the needs of work. The internal bodies under the department are managed by it at its discretion. Its employees can be transferred to other jobs and their salary is determined on its own based on market factors. It has also established an incentive system that links personal emolument with performance appraisal results. These measures have created favorable conditions for attracting high-end informatization talents and improving the capability of talent teams.

As the reform of the new-generation information technology and the government reform are further integrated, e-government construction has ushered in new opportunities. To accelerate e-government construction, it is necessary not only to further strengthen overall planning and coordination, infrastructure construction and new technological application, but also to speed up government business reconstructing and procedure redesigning, improve government functions and better develop and utilize public information resources. This can make digital government construction truly an important approach to stimulate the vitality of economic and social development and help achieve the "Two Centenary" goals.

Chapter 5
Construction and Management of Online Contents

5.1 Outline

In the current era, the new round of technological revolution represented by information technology is bringing profound changes to the communication pattern. The rapid development of Internet is promoting the transmission and sharing of ideas, culture and information on a wider range. The media landscape and public opinion ecosystem are undergoing overall restructuring. The construction and management of online contents is ushering in new major opportunities as well as unprecedented challenges.

Since the 18th National Congress of the CPC, General Secretary Xi Jinping has put forward a series of new ideas, views and judgments on intensifying network ideology and improving the level of Internet governance and application, which are considered the fundamental principles for the construction and management of web contents under the new situation. At the National Propaganda and Ideology Work Conference in August 2018, General Secretary Xi Jinping stressed the importance to scientifically understand the laws of Internet communication, improve the level of Internet governance and application, and turn the Internet's largest variable into the largest increment of career development. In January 2019, General Secretary Xi Jinping led members of the CPC Central Committee Political Bureau to visit *The People's Daily* and presided over the 12th group study session of the CPC Central Committee Political Bureau that was themed on the integrated media development in the omni-media era. He stressed the efforts to apply the achievements in information revolution, boost integrated media development and amplify mainstream tone in public communication so as to consolidate the common theoretical foundation for all Party members and all the people to unite and work hard, and provide great spirit power and public opinion support for realizing the "Two Centenary" goals and the great rejuvenation of the Chinese nation.

Over the past year, under the guidance of the spirit of the important speeches of General Secretary Xi Jinping, the construction and management of web contents has focused on celebrating the 70th anniversary of the founding of the People's Republic

© Publishing House of Electronics Industry 2021

Chinese Academy of Cyberspace Studies, *China Internet Development Report 2019*, https://doi.org/10.1007/978-981-33-6930-6_5

of China and always adhered to the principle of "maintaining positive energy, keeping things under control and correctly utilizing Internet". Besides, the government has kept up with the times and served the expectations of the people through scientifically understanding the laws of Internet communication, improving the level of Internet governance and application, constantly amplifying mainstream tone in online public communication, continuously enhancing the ability of network communication, and steadily promoting the construction of China's network comprehensive governance system. As a result, the wave of positive energy of cyberspace is building and the underlying values hold greater appeal than ever before.

5.2 Continuously Amplifying the Mainstream Tone in Online Public Communication

The year 2019 marks the 70th anniversary of the founding of the People's Republic of China and is critical for achieving the first centenary goal of building a moderately prosperous society in all respects. China has kept to the right path, made innovations and taken proactive actions for online public communication. Relevant parties have thoroughly studied, publicized and implemented Xi Jinping Thought on Socialism with Chinese Characteristics for a New Era. It has focused on celebrating the 70th anniversary of the founding of the People's Republic of China, increased efforts in major theme publicity and agenda setting, and continued to innovate the methods of propaganda, enrich communication carriers and improve the quality of content. In doing so, the transmissibility, leadership, influence and credibility of online news media have been constantly improved, thus creating a favorable atmosphere of online public opinions for celebrating the 70th anniversary of the founding of the People's Republic of China.

5.2.1 Effective and Efficient Online Propaganda of Xi Jinping Thought on Socialism with Chinese Characteristics for a New Era

Xi Jinping Thought on Socialism with Chinese Characteristics for a New Era is the latest achievements of the sinicization of Marxism as well as a guide to the action of the entire Party and all the Chinese people who strive for the great rejuvenation of the Chinese nation. Leading media and business websites have focused on giving play to the leading role of ideology, consciously prioritized the online propaganda of Xi Jinping Thought on Socialism with Chinese Characteristics for a New Era, and interpreted its historical status, spiritual essence, rich connotations, practical demands and era values in an all-round and three-dimensional manner. China's central and local news websites, theory websites and main commercial websites have given full play to the important role of Internet in strengthening ideological guidance and continued to further publicize and interpret Xi Jinping Thought on Socialism with Chinese Characteristics for a New Era. With their efforts, the Party's innovative theories have "flied into the homes of ordinary people" through Internet and the Marxism for contemporary China and the 21st century has become alive and taken root in people's hearts.

Website platforms have stuck to multi-angle interpretation, multi-channel participation and full-platform coverage in their online propaganda work. They have vigorously explored new forms, new channels and new voices of the online propaganda and report of Xi Jinping Thought on Socialism with Chinese Characteristics for a New Era through always complying with the law of propagation and transforming the reporting concepts and methods. Vivid pictures and expressions have aroused strong emotional resonance among Internet users. Major news websites such as people.cn and xinhuanet have continuously strengthened agenda setting and utilized big data to classify their reporting themes, contents and audiences. They have also promoted precision production, intelligent push and interactive communication, and generated a number of interesting original reports. Qstheory.cn and Himalaya FM have jointly produced an audio version of *Thirty Lectures on Xi Jinping Thought on Socialism with Chinese Characteristics for a New Era* named "Understanding Xi Jinping Thought on Socialism with Chinese Characteristics for a New Era in Thirty Days". It was updated on a daily basis and reproduced synchronously by key central news websites and mainstream commercial media. It has caused a great upsurge in online studies with more than 13 million audience within 30 days after its release.

Various websites have given play to their own advantages in consciously telling the China story well, actively engaging in external publicity of Xi Jinping Thought on Socialism with Chinese Characteristics for a New Era, and continuously deepening the global community's understanding and recognition of the China Road, China Model and China Program. CRI Online has carried out *"Xue Xi You Dao"* multilingual theoretical communication campaign, which focuses on interpreting General Secretary Xi Jinping's golden expressions such as reform and opening-up, livelihood economy, and the Belt and Road Initiative, and deeply expounding his new ideas, views and judgments on the country's governance by means of new media such as multilingual animations and short videos in a language that is easy-to-understand and acceptable to foreign Internet users. People.cn has published vivid practices of Xi Jinping Thought on Socialism with Chinese Characteristics for a New Era in many localities of China by means of multimedia reports (articles, images and short videos) on the English channel of People's Daily Online and *People's Daily* client. The reports reflected the great changes of urban and rural areas in China in a three-dimensional, concrete and authentic manner and were forwarded by foreign media such as *Daily Telegraph* in the UK, thus producing great communication effect.

5.2.2   Strong Atmosphere of Online Theme Publicity on Celebrating the 70th Anniversary of the Founding of the People's Republic of China

The 70th anniversary of the founding of the People's Republic of China is an important milestone in the course of China's development. Major network media have consciously kept abreast the trend of the times and satisfied people's expectations. By taking advantage of the chance of celebrating the 70th anniversary of the founding of the People's Republic of China, they have focused on showing the glorious course of New China and vividly presenting the new look of the Chinese people, to ceaselessly arouse the people's patriotism and gather the strength of advance. With deep knowledge in content creation, major central news websites have innovated their publicizing and reporting means, set up special columns and launched a series of publicity reports such as "70 Years of Magnificence · Striving for a New Era" "Patriotic Feeling ·Striver" and "My Motherland and Me", and deeply dug into the stories behind models of the time and advanced figures. "70 Years of Magnificence · Striving for a New Era" reports deeply dug into the stories of ordinary people and ordinary families growing together with the country and presented vivid pictures of the economic development, social progress and people's well-being in representative, landmark and typical regions as well as the earth-shaking changes in various industries after the founding of New China. Besides, in interviews such as "I Love This Blue Country", "Party Flag in Border Areas", "Securing a Decisive Victory in Poverty Alleviation & Securing a Decisive Victory in Building a Moderately Prosperous Society", the reporters went deep into the grass-roots level and the front line to interview the ordinary people.

As a result, a great many of vivid network works fully displaying the glorious history, great achievements and valuable experience made over the past 70 years after the founding of the People's Republic of China as well as the thriving development of the country and concerted and strenuous efforts of the Chinese people have been produced. They have condensed enormous spiritual strength of concerted, pioneering and struggling efforts in developing China. There has appeared also an upsurge of web reports on celebrating the 40th anniversary of the reform and opening-up. The publicity campaign themed on "Reform and Opening-up Starts Again" has produced a great many of creative works that fully display the achievements of comprehensively deepening the reform and opening-up since the 18th National Congress of the CPC from aspects such as history, reality, future, state, society and individuals, aiming to inspire new efforts of the reform and opening-up. According to statistics, more than 400,000 pieces of related works have been released online with the page views exceeding 5.3 billion. Major news websites such as people.cn, xinhuanet and cctv.com have actively engaged in the publicity and propaganda on securing a decisive victory in poverty alleviation and building a moderately prosperous society. They have vividly explained poverty alleviation policies, told typical cases of poverty alleviation and displayed the achievements of poverty alleviation and moving stories of people in impoverished areas making unremitting efforts to get rid of poverty and live a better life. Centering on the commemoration of the 100th anniversary of May Fourth Movement, major news websites and video websites have reported positively on the online campaign of "Youth Singing for Our Motherland · Inheritance" and reproduced relevant videos and reports of Fudan University and other colleges and universities. The Weibo topic "Youth Singing for the Motherland" has triggered active participation and interaction among Internet users and effectively inspired patriotism of the young people. In addition, major network media have set up special columns such as "Steady Progress in China" to talk about hot issues concerning people's livelihood such as consumption, prices, stock market, employment, personal income tax, medical care, education and housing, thus displaying China's political stability, economic development, cultural prosperity, social harmony, good ecology and people's happiness in the new era. Besides, many in-depth and rational articles with powerful data, clear facts and detailed cases have been released to actively respond to the concerns of Internet users and answer their questions, which have further united and reassured the public and enhanced their confidence.

5.2.3 Highlights in the Online Publicity of the Classical Marxism Theory and the Party's Innovative Theory

Through Internet communication and new models of communication, online publicity of relevant theories has continuously produced results, which has made the classical Marxism theory and the Party's innovative theory deeply rooted in people's hearts. In January 2019, the study platform "*Xue Xi Qiang Guo*" was launched nationwide. The platform, consisting of the PC terminal and mobile client, focuses on propagating and implementing Xi Jinping Thought on Socialism with Chinese Characterristics for a New Era and the spirit of the 19th National Congress of the CPC, and contains huge amounts of data of periodicals, ancient books, demonstration lessons, books, songs, operas, films and other materials. It has greatly satisfied the learning needs of the Party members and cadres and the public for mainstream news information under the Internet conditions. As of April 2019, the platform announced more than 100 million registered users, among whom 40–60% were daily active users. It has become a flagship strength to promote the modernization and popularization of the Party's theory. Online theory communication shows such as "*Li Shang Wang Lai*" have further enhanced their influence. Special columns on the themes of "Reform and Opening-up" and "Striving for the New Era" have been set up, the articles on which have been forwarded by large commercial websites and received many hits and favorable reviews.

In order to celebrate the 40th anniversary of reform and opening-up, Qstheory.cn and Toutiao jointly launched the micro-video series of "Ten Key Points to Understand China's Reform and Opening-up", making the communication of the theoretical knowledge of reform and opening-up in the new form of short videos possible. The micro-videos reached more than 500 million audiences within one month after their release, with the total number of hits across the network reaching up to 120 million. This has accumulated valuable experience for innovation in theoretical communication. In order to celebrate the 200th anniversary of Marx's birth, the SMG Radio Centre of Radio and Television Station of Shanghai (RTS), Archimedes FM mobile client, Party School of CPC Shanghai Municipal Committee and other units have jointly launched the online audio program *Telling Stories about Marx to the Post-90 s Generation*. Through telling 19 stories throughout Marx's life, it has helped the young people gain a better understanding of Marx's life journey and the implications of his important ideological achievements on contemporary China. The program has received more than 270 million hits across the network and set off an upsurge for the young people to listen to Marx's life and learn Marxism theory.

5.3 Continuous Improvement of the Internet Communication Capability

General Secretary Xi Jinping points out that Internet is increasingly becoming a new space for people to live and produce and should also become a new space for the Party and government to build consensus. In order to strengthen the ability of network communication and enhance the width and depth of mainstream values, General Secretary Xi Jinping stressed at a group study session of the CPC Central Committee Political Bureau that to promote integrated media development and omni-media construction became an urgent task, and we should take advantage of the fruits of the information revolution to further promote integrated media development. All relevant departments and online media have kept abreast of the mobile-oriented trend of Internet communication and worked hard to effectively integrate various media resources and production factors and combine information content, technological application, platform terminals and management instruments through process optimization, platform reengineering and promoting integrated media development and media transformation and upgrading by utilizing new technologies. As a result, China has continuously expanded the network coverage, gradually optimized the content production mechanism, increasingly diversified network culture products and sped up strengthening its ability of network communication.

5.3.1 Integrated Media Development Continuously Deepening

News media at all levels and of all types have conformed to the current profound changes in media landscape and public opinion ecosystem by strengthening Internet thinking and focusing on shifting from one-way communication to interactive, service-oriented and experiential communication, from single-form communication to omni-media communication (text, image, audio/video), and from single-form communication (newspaper, periodical and radio program) to all-platform communication (websites, microblog, WeChat, electronic bulletin board, mobile phone newspaper and network television). Besides, efforts have also been stepped up in exploring the communication channels for expanding mainstream voices by utilizing new Internet technologies. As of December 31, 2018, 761 online news providing entities received approvals from cyberspace offices of corresponding levels. The services provided by these entities included 743 Internet websites, 563 Apps, 119 forums, 23 blog accounts, 3 Weibo accounts, 2,285 official WeChat accounts, 1 instant messaging tool, 13 livestreaming platforms and 15 other services, accounting for 3,765 services.[1] The omni-media communication landscape has further expanded.

5.3.1.1 Building New Mainstream Media Becomes an Important Development Goal

At present, central mainstream media represented by *People's Daily*, Xinhua News Agency and China Media Group continue to speed up the process of media convergence and institutional reform and seize the commanding heights of new media development strategy through mechanism innovation and department integration. By relying on the "Central Kitchen" platform, *People's Daily* has redesigned production processes, opened internal and external channels, optimized management decisions and innovated business models. It has established dozens of media convergence studios, some of which, including *"Xia Ke Dao"* and *"Xue Xi Da Guo"*, have become very famous. The business scope of *People's Daily* has expanded from newspaper to a "media matrix for people" that involves more than 10 types of media, including newspaper, network, terminal, WeChat and electronic screen, and covers up to 786 million audience in total.[2] Xinhua News Agency has attached great importance to the impact of AI technology on the news industry. It has focused on deeply integrating AI to news scenarios and sped up the construction the world's first intelligent editorial office which utilizes intelligent technology to realize human-machine collaboration and greatly improve the production and communication efficiency. It has also embedded intelligent production technology such as "Media Brain" throughout the production process and comprehensively promoted the application of intelligent technology. With these efforts, it has realized integrated command, multi-link coordination and multi-terminal distribution throughout the processes of planning, collection, editing, contribution and communication, as well as deep integration of technological construction and content construction. The newly established News

---

[1]Licenses to Internet News Providers, January 11, 2019, http://www.cac.gov.cn/2019-01/11/c_1 122842142.htm.

[2]Wang Xiaodong, Bring the Party and the Public Closer—On the Third Anniversary of Xi Jinping's "2·19" Speech, February 19, 2019, see http://he.people.com.cn/n2/2019/0219/c192235-32654181-2.html.

Center of China Media Group has focused on building a clustered, three-dimensional and ecological matrix of news reports, thus greatly improving its ability of omni-media communication. In order to celebrate the 70th anniversary of the founding of the People's Republic of China, the News Center of China Media Group conducted a series of interviewing activities to take journalists to retrace the route of the Long March. They included the special program "Restart the New Long March" in the news channel and the "Journalists Retrace the Route of the Long March" program in the "Voice of China" channel. Special columns were also set up by the New Media Department of CCTV News to broadcast live images and videos. They have received wide attention among Internet users.[3]

5.3.1.2 Deep Media Convergence Provides a New Platform for Communicating with, Guiding and Serving the Masses

Local key news websites have given full play to their natural advantages of being close to the masses at the grass-roots level, constantly expanded the management concepts and innovated relevant system and mechanism. They have utilized their own network platforms and commercial network platforms to amplify the communication effect, and gradually built a great number of first-class dynamic network fields with local characteristics, which have provided important platforms and channels for listening to the voices of the public and reporting grass-roots work, as well as great power and support for publicizing the mainstream voice and purifying the network ecosystem. For example, ThePaper.cn has transformed from a phenomenal network new media product into a platform-level new mainstream Internet media over the past five years. As of the end of 2018, the downloads of the mobile app of ThePaper.cn reached 110 million, with 9.5 million daily active users. It has become a major force to lead local new media development and enjoyed higher visibility and influence in China and abroad.

5.3.1.3 County-Level Convergence Media Centers Become a New Area for Integrated Media Development

In 2019, the Publicity Department of the CPC and the National Radio and Tele-vision Administration (NRTA) jointly issued a number of policy and specifica-tion documents, including *Construction Specifications of County-Level Converged Media Centers* and *Cybersecurity Specifications on County-Level Converged Media Centers*, to propose general requirements and make clear deployment for the construction of county-level converged media centers. All localities have actively built county-level converged media centers to integrate county-level media resources,

---

[3]Li Zhi, Ten Keywords of the Deep Media Convergence in China Media Group, August 15, 2019, China Society of Motion Picture and Television Engineers, see http://www.ttacc.net/a/news/2019/0815/57925_2.html.

adjust and optimize media layout and strengthen the role of mainstream public opinions. Besides, they have expanded government services and life services and focused on building information services complexes such as "media + government affairs", "media + services", "media + e-commerce" and "media + cultural innovation". Some regions have developed their own distinctive models for the construction of county-level converged media centers based on local conditions. For example, based on the model of group operation, Changxing County of Zhejiang Province has promoted in-depth omni-media integration in aspects such as content, channels, platforms, operation and management, and innovated the developed the content of "news services + government services". Besides, it has realized government information sharing and business collaboration through cloud computing data centers and comprehensive grassroots governance platforms. As the result, the goal of "letting data run more and let the masses run less" has been truly achieved.[4] Fenyi County of Jiangxi Province has developed the "*hua ping fen yi*" client based on its county-level converged media center, which integrates 7 types of media ports, namely, the broadcasting stations, TV stations, newspapers, government microblogs, WeChat official accounts, mobile newspapers and government websites. Besides, it has set up a series of columns such as "Party and Masses", "Township Affairs", "Village Community" and "Seeking Public Opinions on Governance", launched many convenience services and implemented the function of integrating news, government affairs and services.

5.3.1.4 Government New Media Are Flourishing

In December 2018, the General Office of the State Council released *Opinions on Promoting the Sound and Orderly Development of Government New Media* (GBF [2018] No. 123), stressing that government new media is an important channel for the CPC and the government to engage with, serve, and bond with the public in the era of mobile Internet, a significant means to accelerate the transformation of government functions and build a service-oriented government, a consequential position to guide online public opinion and create a clean cyberspace, and a major method to explore new models of social governance and improve social governance capacity. Besides, all regions and departments should build high-quality government new media and use them well. The government departments at all levels have actively developed government new media based on Internet platforms. As of June 2019, 31 provinces (autonomous regions, municipalities directly under the Central Government) had their own accounts of WeChat City Services, government microblog accounts and government Toutiao accounts. Among them, WeChat City Services had a total of 620 million users, the number of verified government accounts on Sina Weibo reached

---

[4]"Changxing's Exploration" of the Development of County-Level Media Convergence, *Guangming Daily*, December 7, 2018, see http://epaper.gmw.cn/gmrb/html/2018-12/07/nw.D110000gmrb_201 81207_1-07.htm.

139,000, and the number of Toutiao accounts opened by governments at all levels reached 81,168.[5]

5.3.2 Continuous Optimization of Content Production Models

The mainstream media have actively grasped the new changes in the communication rules in the Internet age and kept up with the new characteristics of the content consumption demands of Internet users. During online publicity and report activities, they focus on giving play to their advantages in content, implement the excellent works strategy and continue to report on new business formats in the new era with new voices and expressions. A large number of excellent online news works have been produced, resulting in further improvement of the arrival rate, reading rate and liking rate of online news information. In November 2018, winners of the 28th China News Award were announced, with 348 pieces of works winning awards. Among them, five were granted the Special Prize, including one web design work and one converged media broadcast work, and 62 were granted the First Prize, among which there were 15 or nearly 25% involving web reviews, special reports on web pages, short videos of converged media, columns of converged media, innovation in converged media and other topics. They have received wide recognition by all circles of the society. See Table 5.1 for the winning works in web contents of the 28th China News Award (Special Prize and First Prize).

Thanks to the increasing diversity in network platforms and the popularization of creation and production, the UGC (user generated content) model and PGC (professional generated content) model for web content production have further developed into the MCN (Multi-Channel Network) model that links content producers, platforms and advertisers. Social media has become a cutting-edge platform for exploring the model of web content production. The microblog content was independently created and released by users in early stages, which later became more professional as specialized content producers opened microblog accounts. At present, the content production model of microblogs is undergoing a transition to MCN, which can produce high-quality content in a sustained and stable manner by combining PGC and capital operation. MCN has also become a trend for short video creation. In 2019, Kwai launched the "Photosynthetic Plan" to help foster 100,000 high-quality creators. TikTok also launched the "Blue V Ecology Project" to make the content production of short videos more specialized. With the combination of UGC, PGC and MCN production models, web content production has become more creative and dynamic and more high-quality content is produced constantly. Faced with the new development trend, mainstream media and business platforms have explored new

---

[5]The 44th "Statistical Report on China Internet Development", China Internet Network Information Center (CNNIC), August 30, 2019, see http://www.cnnic.net.cn/hlwfzyj/hlwxzbg/hlwtjbg/201908/t20190830_70800.htm.

Table 5.1 Winning Works in Web Contents of the 28th China News Award (Special Prize and First Prize)[a]

Number	Prize	Item	Title	Publishing unit/(publishing account
1	Special Prize	Web design	Matrix Design of CCTV Zero-Homepage Report on the 19th National Congress of the CPC	cctv.com
2	Special Prize	Converged media broadcast	Two Sessions Underway	Weibo account "*ren min wang fa ren*"
3	First Prize	Web reviews	Disgusting! Use Comfort Women's Headshots as Internet Memes, Where is the Conscience?	cyol.com
4	First Prize	Online report	"Way to the Paradise" on the Cliff	cqnews.net
5	First Prize	Online report	Original Aspiration	CCTV News Client
6	First Prize	Online interview	Authoritative Experts' Analysis on the Underlying Reason for India's Illegal Invasion of Chinese Territory	China News
7	First Prize	Web design	Never Forget the History, Resolutely Carry out Rejuvenation - National Memorial Day for Nanjing Massacre Victims	gdtv.cn
8	First Prize	Converged media short video	A Breakout Story in Rongshui, Liuzhou ǀ A Reporter of *Guangxi Daily* Lost Contact for Dozens of Hours and Sent back the Latest Images after Passing through 40 Landslides!	*Guangxi Daily* WeChat official account, *Guangxi Daily* client (Guangxi Cloud client)
9	First Prize	Converged media short video	Road of Public Servants	cbox.cntv.cn
10	First Prize	Converged media broadcast	VR Panoramic Live Broadcast of Tianzhou-1's Launch Mission	cbox.cntv.cn, v.qq.com

(continued)

Table 5.1 (continued)

Number	Prize	Item	Title	Publishing unit/(publishing account
11	First Prize	Converged media interaction	"Military Uniform Pictures" H5	People's Daily client
12	First Prize	Converged media interaction	Thumbs-up for the 19th National Congress of the CPC, China is Growing Stronger	Xinhua News client
13	First Prize	Converged media column	Xia Ke Dao	WeChat official account
14	First Prize	Converged media column	Commentaries on International Affairs	WeChat platform
15	First Prize	Converged media interface	Long Interactive Comic Strip ǀ Zunyi's Village Secretary Huang Dafa Spent 36 Years Drawing Water and Digging Canals	ThePaper.cn
16	First Prize	Converged media innovation	The Pilot	Xinhua News client
17	First Prize	Converged media innovation	H5 Report on the "CCTV Anchors' WeChat Moments" Series	"Voice of China" WeChat official account

[a]Catalogue of Award-Winning Works of the 28th China News Award, xinhuanet, November 3, 2018, see http://www.xinhuanet.com/2018-11/03/c_1123656210.htm

cooperation models, with the former utilizing their advantages in content production capacity and social credibility to achieve integration, fusion and innovation of media resources with the latter. Besides, by giving play to the role of commercial platforms as the information distribution channel, they have realized cooperative sharing of information, expansion of propagation range and influence improvement. For example, xinhuanet has published interpretative articles on the major activities of the central government based on its content advantages. Its high-quality content has been continuously tracked and concerned by microblogs, WeChat official accounts, Toutiao, Sohu, NetEase and other major platforms. Thanks to the spread of new media platforms, each report has been reproduced by over 400 media on average.

5.3.3 Brand-New and Diversified Internet Cultural Products

Internet has become an important carrier to carry forward the Chinese outstanding culture and enrich people's spiritual world. Internet culture construction aims to nourish the people and society by promoting the core socialist values and the outstanding achievements of human civilization, and focuses on fostering healthy and positive Internet culture. Over the past year, the forms of Internet culture production have been constantly innovated and enriched, which has not only promoted the prosperity of socialist culture with Chinese characteristics, but also provided strong support for keeping up with people's ever-growing spiritual and cultural needs.

The excellent Chinese culture and core socialist values are spreading well through networks. China's efforts in Internet culture production have focused on building high-quality products, refining the spiritual traits of the Chinese splendid culture and displaying its connotations and essence. Internet has played an important role in spreading Chinese culture and showcasing China's style in the new era. Solid progress has been made in the key programs of the Project for the New Media Communication of Chinese Culture. New media works on the themes of traditional festivals and customs, cultural connotation and homestead feelings were carefully orchestrated during traditional festivals such as the Spring Festival, Lantern Festival, Dragon Boat Festival, Mid-autumn Festival and Double Ninth Festival. The program "Our Festivals" created a festive atmosphere of reunion, civility and harmony on Internet through intensified online communication efforts, such as multi-connotation interpretation, three-dimensional dissemination and all-round display. Other online communication programs such as "Ode to China's Cultural Tradition · Thousand Years in One Webpage", "Ode to China's Cultural Tradition · Famous Masters" and "Ode to China's Cultural Tradition · Intangible Cultural Heritage" have also effectively promoted the creative transformation and innovative development of Chinese excellent traditional culture and made it get alive and popular. The core socialist values are effectively disseminated on Internet. Major central news websites and commercial websites have delved into the connotations and typical cases of the core socialist values, and painstakingly built a series of branded programs such as "Practitioners of the Chinese Dream" through multi-media means that incorporates images, text, video and H5 to tell the stories of Chinese people striving hard to realize their dreams, and to guide the Internet users, particularly the young people, to actively practice the core socialist values and spread positive energy. In order to carry forward the Chinese spirit, including the Red Boat spirit, Long March spirit, Hongyan spirit, the spirit of manned space flight and the spirit of reform and opening-up, network media have carried out an online publicity event themed on "Power of Spirit · Soul of the New Era" to showcase the vivid practices of various regions and discuss the epochal connotations of the Chinese spirit. A great many of network high-quality columns have been produced and the page views of relevant topics have reached several billion.

As Internet culture industry is maturing and its order is gradually standardized, various Internet cultural products such as online music, online games, online videos and online broadcast have been booming. With continuous efforts in online videos, music Apps represented by Tencent Music and NetEase Cloud Music have closed financing and promoted the rapid development of online music. As of June 2019, the user size of online music was 608 million, up 32.29 million over the end of 2018.[6] Online games have seen steady development. As of June 2019, the user size of online games was 494 million, up 9.72 million over the end of 2018; the number of mobile game users was 468 million, up 8.77 million from the end of 2018.[7] As domestic game operators such as Tencent, NetEase and Perfect World established cooperation with foreign game developers such as Ubisoft and Valve, domestic games have started to go global. Online video has seen stable growth. Up to June 2019, the user size of online video was 759 million or 88.8% of China's total netizen population, up 33.91 million over the end of 2018.[8] Short videos have been developing fast in particular. As of June 2019, there were 648 million short video users, making up 75.8% of China's Internet users.[9] In the first half of 2019, the average time of short video exceeded 22 h, up 8.6% year on year.[10]

5.4 Continuous Improvement of the Integrated Internet Governance System

The construction of a system for integrated Internet governance is an important strategic decision made at the 19th National Congress of the CPC. In July 2017, *Opinions on Speeding up the Establishment of a Comprehensive Internet Governance System was deliberated and passed by the Central* Comprehensively Deepening Reforms Commission, which clarified the guiding ideology, basic principles, development goals and main tasks of the construction of the Internet governance system, and further improved the top-level design of the system. All regions and all departments have solidly pushed forward the construction of the comprehensive Internet

---

[6]The 44th "Statistical Report on China Internet Development", China Internet Network Information Center (CNNIC), August 30, 2019, see http://www.cnnic.net.cn/hlwfzyj/hlwxzbg/hlwtjbg/201908/t20190830_70800.htm.

[7]The 44th "Statistical Report on China Internet Development", China Internet Network Information Center (CNNIC), August 30, 2019, see http://www.cnnic.net.cn/hlwfzyj/hlwxzbg/hlwtjbg/201908/t20190830_70800.htm.

[8]The 44th "Statistical Report on China Internet Development", China Internet Network Information Center (CNNIC), August 30, 2019, see http://www.cnnic.net.cn/hlwfzyj/hlwxzbg/hlwtjbg/201908/t20190830_70800.htm.

[9]The 44th "Statistical Report on China Internet Development", China Internet Network Information Center (CNNIC), August 30, 2019, see http://www.cnnic.net.cn/hlwfzyj/hlwxzbg/hlwtjbg/201908/t20190830_70800.htm.

[10]*Semi-Annual Report on the Short-Video Industry in 2019*, QuestMobile, July 22, 2019, see https://www.questmobile.com.cn/research/report-new/58.

governance system, constantly improved relevant laws and regulations, and continuously enhanced the capacity of using technology in Internet governance. Besides, as Internet platforms have continued implementing their main responsibilities, the comprehensive Internet governance has further produced positive effects.

5.4.1 Continuous Improvement of the Leadership and Command System

In March 2018, the CPC Central Committee issued *Plan for Deepening the Reform of Party and Government Institutions*. As a result, the original Cyberspace Affairs Leading Group was transformed into the Central Cyberspace Affairs Commission, and the duties of the Office of the Central Cyberspace Affairs Commission were optimized. Meanwhile, the management agency of National Computer Network Emergency Response Technical Team/Coordination Center of China was changed from the Ministry of Industry and Information Technology (MIIT) to the Office of the Central Cyberspace Affairs Commission. Over the past year, by taking the advantage of the valuable opportunity of China's institutional reform, the cyberspace affairs management systems at the three levels of the central government, provinces (autonomous regions, municipalities directly under the Central Government) and cities have been further improved. Besides, all provinces (autonomous regions, municipalities directly under the Central Government) have established provincial cyberspace affairs commissions and corresponding offices. Some of them have established cyberspace affairs commissions at the county level. The national cyberspace affairs system has actively promoted the establishment of relevant working mechanisms based on the function orientation and mission requirements of cyberspace affairs departments. Through establishing and improving the systems for overall planning and coordination, emergency management, negotiation over major issues, supervision of major decisions and disclosure of essential information, the mechanisms for exercising and supervising Internet governance and management power have been continuously standardized, which has ensured the full implementation of the territorial management responsibilities. Zhejiang, Guangdong, Shandong and other provinces have actively explored to establish the mechanisms for collaboration on comprehensive web content governance and focused on improving the mechanisms for inter-departmental, cross-level, cross-regional, cross-system and cross-business division and cooperation of web content governance. The mechanisms have achieved efficient operation through coordination between government regulation and Internet user's self-discipline, content security and content innovation, user management and platform management, and the systems and mechanisms for Internet management have been further improved.

All regions and all departments have conscientiously implemented the system of responsibility for network ideological work, and formulated and issued detailed working measures or implementation plans to further clarify the responsibilities of Party members and cadres, especially leading cadres, for the ideological work. Besides, they have been resolutely safeguarding the network "responsibility fields". As a result, the system of responsibility for network ideological work has been gradually implemented. For example, some provinces, including Shanxi, have issued *Implementation Opinions on Further Strengthening the Cybersecurity affairs across the Province*. It arranges and plans the important task of implementing the system of responsibility for network ideological work, and puts forward clear instructions for the competency standards, training and learning, and appraisal, reward and punishment and organizational guarantee for the leading cadres' network ecosystem governance work, ensuring that work duties are performed by relevant departments.

5.4.2 Continuous Deepening of Law-Based Internet Governance

The rule of law is a long-acting and fundamental means, and to improve the legal system of Internet governance is a key step in the construction of the comprehensive Internet governance system. With the rapid development of Internet every day, new situations and problems are emerging endlessly, so that relevant laws and regulations are urgently needed for regulation and guidance. Facing the outstanding issues in Internet development, the National Internet Information Office has done a lot of legislative work, including issuing more than 20 documents of departmental regulations and management regulations, and particularly standardizing instant messaging tools, web broadcast, online audio-visual programs, applications, public accounts, groups, comments on posts, etc. In January and August 2019, the Office issued *Provisions on the Administration of Blockchain Information Services* and *Provisions on the Internet Protection of Children's Personal Information* respectively. In September 2019, it solicited public opinions on *Provisions on Network Ecological Governance* to further improve relevant laws and regulations on comprehensive Internet governance. The Publicity Department of the CPC, the Ministry of Education, the Ministry of Culture and the National Radio and Television Administration (NRTA) have also issued a series of administration opinions and regulations based on their own functions, to provide strong legal support for law-based Internet governance.

In order to create a clean and sound Internet environment, the cyberspace affairs departments at all levels have actively implemented their network law enforcement duties and strengthened network law enforcement by applying more severe punishment. The Office of the Central Cyberspace Affairs Commission has launched a 6-month (starting from January 2019) special action of network ecosystem governance across the country jointly with relevant departments, which aims to remove 12 types of negative and harmful information (namely, pornographic information, vulgar information, violent information, horrifying information, gambling and fraud information, Internet rumors, feudalistic superstition information, abusing and spoofing information, clickbait information, threatening information, hate mongering information, and the information that spreads unhealthy lifestyles and unhealthy popular culture) and effectively contain the backlash of various harmful information. The action has produced remarkable effects. For example, a special action was launched against the chaos in online video industry. Punitive measures such as verbal warning, unshelving and service shutdown were taken according to relevant laws and regulations against 26 illegal audio platforms that disseminated historical nihilism and pornographic information.[11] The cyberspace affairs departments in Beijing, Shanghai, Tianjin and other regions have actively performed their territorial management duties, strengthened governance and law enforcement for network ecosystem, carried out ongoing renovation activities, and constantly regulated the order of online communication. As of July 2019, China's cyberspace affairs departments had summoned 1,333 websites, given warning to 884 websites and ordered 181 websites to pause updates. Besides, 4,986 illegal websites had their web licenses and filings removed or were shut down by cyberspace affairs departments and telecommunication administrations, and 663 related cases were transferred to judicial departments. According to user service agreements, relevant websites closed nearly 300,000 illegal accounts and groups.[12] See Table 5.2 for relevant regulations and policies on web content governance.

5.4.3   Enterprises Continuing to Fulfill Their Primary Responsibilities

China's Internet enterprises have conscientiously fulfilled the main responsibilities of Internet platforms, placed great emphasis on content management and supervision of the platforms, and strengthened content governance through various means, including improving the technological capacity, perfecting rules and regulations, and regulating process management. In the first half of 2019, with the help of AI technology, Baidu reviewed and disposed of 31.25 billion pieces of harmful information in 11 types, including obscene information, drug information, gambling information, fraud information and infringement information.[13] The WeChat public platform released a series of documents, including *Rules for Complaints Collegiate of "Article Spinning" on the WeChat Public Platform*, aiming to safeguard the health of the content on the

---

[11]The Office of the Central Cyberspace Affairs Commission Focused on Carrying out Special Campaigns to Improve the Order of Online Videos [EB/OL], June 28, 2019, see http://www.cac.gov.cn/2019-06/28/c_1124685210.htm.

[12]China's Administrative Law Enforcement Work of Cyberspace Affairs in the First Quarter Produced New Achievements, April 24, 2019, see http://www.cac.Gov.cn/2019-04/24/c_1124410176.htm; China's Administrative Law Enforcement Work of Cyberspace Affairs in the Second Quarter Forged ahead Constantly, July 29, 2019, see http://www.cac.gov.cn/2019-07/29/c_1124812129.htm.

[13]Baidu's Semi-Annual Report on Information Security Governance, Beijing News, July 11, 2019, see http://www.bjnews.com.cn/finance/2019/07/11/602394.html.

Table 5.2 Regulations and Policies on Web Content Governance

Number	Time	Issuing agency	Title	Key provisions
1	August 2014	National Internet Information Office	*Interim Provisions on the Administration of the Development of Public Information Services Provided Through Instant Messaging Tools*	Promoting the healthy and orderly development of public information services provided through instant messaging tool, protecting the legitimate rights and interests of citizens, legal persons and other organizations, and safeguarding national security and public interests
2	February 2015	National Internet Information Office	*Provisions on the Administration of Account Names of Internet Users*	Strengthening the administration of account names of Internet users, and protecting the legitimate rights and interests of citizens, legal persons and other organizations
3	June 2016	National Internet Information Office	*Provisions on the Administration of Internet Information Search Services*	Regulating Internet information search services, promoting the healthy and orderly development of Internet information search industry, protecting the legitimate rights and interests of citizens, legal persons and other organizations, and safeguarding national security and public interests
4	June 2016	National Internet Information Office	*Provisions on the Administration of Mobile Internet Applications Information Services*	Strengthening the administration of mobile Internet applications information services, protecting the legitimate rights and interests of citizens, legal persons and other organizations, and safeguarding national security and public interests

(continued)

Table 5.2 (continued)

Number	Time	Issuing agency	Title	Key provisions
5	July 2016	National Internet Information Office	*Notice on Further Strengthening Management and Prohibition of False News*	Combating and preventing online false news and regulating the information dissemination order of online news
6	November 2016	National Internet Information Office	*Provisions on the Administration of Internet Live-Streaming Services*	Strengthening the administration of Internet live-streaming services, protecting the legitimate rights and interests of citizens, legal persons and other organizations, and safeguarding national security and public interests
7	May 2017	National Internet Information Office	*Provisions for the Administration of Internet News Information Services*	Strengthening the administration of Internet news information services, and promoting the healthy and orderly development of Internet news information services
8	May 2017	National Internet Information Office	*Provisions on the Administrative Law Enforcement Procedures for Internet Information Content Management*	Regulating and ensuring that Internet information content management departments perform their responsibilities of administrative law enforcement by law and impose administrative penalties appropriately, and promoting the healthy and orderly development of Internet information services
9	July 2017	National Internet Information Office	*Notice on Registration and Archival Filing of National Internet Live-Streaming Service Enterprises*	Intensifying rectification of live-streaming chaos
10	August 2017	National Internet Information Office	*Provisions on the Administration of Internet Comments Posting Services*	Regulating Internet comments posting services

(continued)

Table 5.2 (continued)

Number	Time	Issuing agency	Title	Key provisions
11	August 2017	National Internet Information Office	*Provisions on the Administration of Internet Forum and Community Services*	Regulating Internet forum and community services, promoting the healthy and orderly development of the Internet forum and community industry, protecting the legitimate rights and interests of citizens, legal persons and other organizations, and safeguarding national security and public interests
12	September 2017	National Internet Information Office	*Provisions on the Administration of Internet User Public Account Information Services and Provisions on the Administration of Internet Group Information Services*	Regulating the administration of Internet user public accounts and various Internet groups
13	October 2017	National Internet Information Office	*Provisions on the Administration of the Safety Assessment of New Technologies and Applications for Internet News Information Services*	Regulating the safety assessment of new technologies and applications for Internet news information services, safeguarding national security and public interests, and protecting the legitimate rights and interests of citizens, legal persons and other organizations

(continued)

Table 5.2 (continued)

Number	Time	Issuing agency	Title	Key provisions
14	October 2017	National Internet Information Office	*Measures for the Administration of Content Management Practitioners Working for Internet News Information Service Providers*	Strengthening the administration of content management practitioners working for Internet news information service providers, protecting the legitimate rights and interests of practitioners and social public, and promoting the healthy and orderly development of Internet news information services
15	December 2017	Eight ministries (including the Publicity Department of the CPC)	*Opinions on Strictly Regulating the Administration of the Online Gaming Market*	Focusing on rectifying illegal acts and inappropriate content of online games
16	February 2018	National Internet Information Office	*Provisions on the Administration of Micro-blogging Information Services*	Promoting the healthy and orderly development of micro-blogging information services, protecting the legitimate rights and interests of citizens, legal persons and other organizations, and safeguarding national security and public interests
17	March 2018	The National Radio and Television Administration (NRTA)	*Notice on Further Regulating the Order of Online Communications for Internet-based Audio-visual Programs*	Addressing the issues of irregularities in the production and broadcasting of Internet-based audio-visual programs, and regulating the order of communication of Internet-based audio-visual programs

(continued)

Table 5.2 (continued)

Number	Time	Issuing agency	Title	Key provisions
18	November 2018	National Internet Information Office	*Provisions on the Security Assessment of Internet Information Services with Public Opinion Attributes or Social Mobilization Capabilities*	Strengthening the security management of Internet information services with public opinion attributes or social mobilization capabilities and related new technologies and new applications
19	January 2019	National Internet Information Office	*Provisions on the Administration of Blockchain Information Services*	Regulating blockchain information service activities
20	January 2019	China Netcasting Services Association (CNSA)	*Norms for the Administration of Online Short Video Platforms and Detailed Implementation Rules for Online Short Video Content Review Standards*	Regulating the short video industry in aspects of platform management and content review
21	August 2019	National Internet Information Office	*Provisions on the Cyber Protection of Children's Personal Information*	Focusing on the cyber protection of children's personal information and supporting the sound development of children

public platform through standardized systems. Toutiao has constantly strengthened the construction of manual review capabilities, and its review team now has more than 10,000 members. In addition, in order to combat chaos in Internet reproducing and protect media copyright, many Internet enterprises have taken active measures to prevent and control copyright infringement behaviors, such as providing ownership authentication and evidence collection services for original news works through blockchain, public key encryption and trusted-timestamping technologies.[14]

Internet enterprises continue to strengthen industrial self-discipline, and actively participate in building industrial organizations and publishing self-discipline conventions. In December 2018, Tencent, Baidu, iQIYI, Sohu, Sina, Kwai and other companies jointly established the China Network Copyright Industry Appliance and released *Self-discipline Convention for the Copyright of Online Short Videos in China* on the Fifth Symposium of Internet New Copyright Issues in China,[15] which promoted the establishment of the industrial consensus and further standardized the industrial order.

5.4.4 Social Supervision Channels Becoming More Smooth

In recent years, various network platforms in China have gradually established and improved reporting institutions and unblocked the reporting channels to accept and deal with the public's reports of harmful information, such as pornographic, vulgar, rumors, gambling, violent and terrorist information, and make up for the issue of insufficient technological and personnel support in network platform governance. The Center for Reporting Illegal and Unhealthy Information of the National Internet Information Office has opened various reporting channels, including official websites, online reporting apps and "12377" reporting hotline. Besides, it has organized six groups of more than 2,600 websites to announce their reporting channels, making it more convenient for Internet users to report to them. In 2019, China's network reporting departments at all levels accepted a total of 94.039 million cases of Internet illegal and unhealthy information, indicating continuous improvement of the whole society's initiative in jointly participating in cyberspace governance.[16]

The National Internet Information Office and 27 guiding units such as the Ministry of Education and the Ministry of Science and Technology (MOST) have jointly established China's Internet Anti-Rumor Platform, which serves as an authoritative platform for the masses to identify and report rumors as well as a "firewall" to refute rumors through timely verification, active clarification and authoritative denial of

---

[14] Annual Report on China's Network Copyrights Protection 2018, the China Academy of Information and Communications Technology, April 26, 2019, see http://www.199it.com/archives/869531.html.

[15] Internet Enterprises Released the Self-Discipline Convention on Short Video Copyrights, CMG, December 21, 2018, see http://media.people.com.cn/n1/2018/1221/c14677-30481141.html.

[16] As of August 2019. Source: The Center for Reporting Illegal and Unhealthy Information of the National Internet Information Office, see http://www.12377.cn/node_543837.htm.

Internet rumors that endanger national security, disrupt social order, damage the rights and interests of the masses and mislead public opinions. Since its launch in August 2018, the Internet Joint Anti-Rumor Platform has published and posted more than 4,000 anti-rumor articles, and transferred the rumor reporting data of the National Network Reporting Management System and the rumor clues provided by Sina Weibo and other member units to the platform's database. In the first half of 2019, a total of more than 70,000 pieces of anti-rumor information were integrated into the platform, indicating that effects of the Internet joint anti-rumor mechanism have come into play.

5.4.5 Self-discipline Consciousness of Internet Users Continuously Growing

Internet users are the subject of cyberspace. At present, China's Internet users have a generally high level of education and patriotism and their cybersecurity concept and ability to distinguish harmful information continue to strengthen. Many Internet users realize that cyberspace is not a place beyond the rule of law and consciously resist all kinds of illegal and criminal activities such as Internet rumors, terrorism, obscenity, drug trafficking, money laundering, gambling, theft and fraud. The construction of the network credibility system has progressed gradually. The network spirit of "not following the trend blindly" has started to take root in the hearts of Internet users, and the number of Internet users who have the courage, chance and skill to express their opinions and consciously act as the maintainer of good order in cyberspace is increasing.

In February 2019, the 2019 Meeting on Promoting the Project of Striving to Be a Good Chinese Netizen in Beijing was convened by the National Internet Information Office and other departments such as the Ministry of Education, the People's Bank of China, the All China Federation of Trade Union (ACFTU), the Central Committee of the Communist Young League and the All-China Women's Federation (ACWF). The meeting stressed the importance to give play to the positive role of Internet users, deepen network literacy education, improve the mechanism of multi-agent cooperation governance, promote the brand building for Internet cultural activities, attract and engage more Internet users to positively express their voices on Internet, and accelerate the establishment and improvement of the comprehensive Internet governance system. During the Spring Festival of 2019, people.cn and cyol.com launched a campaign to encourage Chinese Internet users to post pictures of their hometowns during the festival. In the campaign, a total of more than 2,000 pieces of works were collected, which displayed the grass-roots construction achievements and people's good quality of life and spread positive energy in different forms.[17] In August 2019, the Third "China's Young Good Netizen" Campaign was launched.

---

[17] Award-winning Works in China's Good Netizen Spring Festival Collection Campaign Were Published, April 11, 2019, see http://www.people.com.cn/n1/2019/0411/c347407-31025145.html.

During the event, a total of 2,095 stories were collected, among which 100 excellent stories were selected and won the applause of the broad netizens.[18]

Cyberspace has become the spiritual home of hundreds of millions of people. At present, the rapid development of the new-generation information technology represented by 5G, AI and big data is bringing new opportunities for web content construction as well as new challenges for web content management. In order to firmly take on the high ground of public opinion guidance, ideological leadership, cultural inheritance and service for people, and create a clean and sound cyberspace, it is imperative to adhere to correct political direction, public opinion orientation and value orientation, constantly boost innovation in online communication concept, content, form, methods and means, continuously promote integrated media development, accelerate the improvement of comprehensive Internet governance capacity, and truly consolidate and strengthen the position for public opinion propaganda of mainstream ideas.

---

[18]Excellent Stories of the Third "China's Young Good Netizen" Campaign Were Unveiled, August 27, 2019, see http://zjnews.china.com.cn/2019hwm/jj/2019-08-27/185623.html.

Chapter 6
Cybersecurity Protection and Assurance

6.1 Outline

There will be no national security without cybersecurity. Cybersecurity is essential for national security and social stability as well as the vital interests of the people. It is increasingly becoming a major issue of overall importance. Currently, China's cybersecurity threats are characterized with extensiveness, diversity and strong concealment. These threats and those in other fields are deeply combined and excite each other, which has caused expansion of national security boundary and made security issues more comprehensive, linked and variable. In view of the severe and complicated cybersecurity situation, China has comprehensively implemented the holistic approach to national security, established correct cybersecurity concepts, ensured both development and security, solidly pushed forward the cybersecurity work, and focused on improving its capacity to protect cybersecurity and strengthening the national cybersecurity barrier.

China's overall cybersecurity situation is severe and complicated. Traditional cybersecurity threats should not be ignored. Vulnerabilities in basic, open source, and application software/hardware such as CPU chips and middleware pose a serious threat to cybersecurity. The frequency of distributed denial-of-service (DDoS) attacks has decreased, but their peak traffic continues to increase. Attacks such as advanced persistent threats (APTs) against national key industrial units occur frequently, and ransomware threats are increasing. The issue of large-scale leakage of user personal information remains severe. As new cybersecurity threats continue to emerge, fake mobile applications become a new channel of Internet fraud, cloud platforms become an important target of cyberattacks, and networked intelligent devices face malicious attacks. Cybersecurity incidents occur frequently in key areas. Industrial Internet, with serious security risks, is undergoing increasing attacks. Internet financial platforms and mobile applications are facing severe security risks. Important information systems and data in the healthcare industry are facing severe cybersecurity challenges. Critical information infrastructure in power and other industries have become a target of attack at the national level.

© Publishing House of Electronics Industry 2021
Chinese Academy of Cyberspace Studies, *China Internet Development Report 2019*, https://doi.org/10.1007/978-981-33-6930-6_6

Solid progress is made in all the aspects of cybersecurity. Positive progress is made in the cybersecurity protection and assurance work. Efforts are stepped up in the security protection of critical information infrastructure, classified protection of cybersecurity, data security administration, personal information protection and security evaluation for cloud computing services. The management and special administration of cybersecurity incidents are further deepened. The cybersecurity industry and technology undergo steady development. Continuous progress is made in the cultivation of cybersecurity professionals. The citizens' protection awareness and skills of cybersecurity are improved constantly.

6.2   Severe and Complicated Cybersecurity Situation in China

China is currently confronted with severe cybersecurity circumstances, increasingly complicated cybersecurity environment and intensified cybersecurity risks.

(1) Traditional cybersecurity threats remain prominent. Vulnerabilities of basic software/hardware are exposed constantly. The peak traffic of denial-of-service (DDoS) attacks continues to increase. Advanced persistent threats (APTs) continue to get worse. Ransomware attacks occur frequently and its variants are increasing. The issue of leakage of personal information remains prominent.
(2) While enriching people's digital life, new technologies and applications such as mobile Internet, cloud platforms and networked smart devices also expand network exposure and bring new threats and risks.
(3) Critical information infrastructure in industrial Internet, Internet financial platforms, healthcare, power and other industries are facing more security problems.

6.2.1   Traditional Cybersecurity Threats Remaining Severe

6.2.1.1   Vulnerabilities in Basic Software/Hardware Such as CPU Chips and Middleware Seriously Threatening Cybersecurity

Cybersecurity vulnerability is an important source of cyber threats and the cause of most cybersecurity incidents. Since 2018, many widely used software and hardware have been disclosed to have security vulnerabilities, which are hard to fix and bring severe challenges to cybersecurity. Examples of hardware vulnerabilities include Meltdown[1] and Spectre[2] CPU vulnerabilities in Intel x86 and x64 hardware. These

---

[1]Meltdown vulnerability: CNVD-2018-00303 corresponding to CVE-2017-5754.

[2]Spectre vulnerability: CNVD-2018-00302 and CNVD-2018-00304 corresponding to CVE-2017-5715 and CVE-2017-5753 respectively.

vulnerabilities affect not only Intel, AMD, ARM processor chips produced after 1995, but also Windows, Linux Mac OS, Android and other operating systems that use the above-mentioned CPU chips as well as the cloud computing facilities of Amazon, Microsoft, Google, Tencent Cloud and Alibaba Cloud and other companies. With these vulnerabilities, attackers can bypass the security isolation mechanism of memory access and obtain the protected data of operating systems and other programs through malicious programs, thus resulting in the leakage of sensitive information from memory. Software vulnerabilities are found to exist in WinRAR, Microsoft Remote Desktop, Oracle Weblogic server, Cisco Smart Install and other basic software, which are widely applied into China's basic applications and general software/hardware products. These vulnerabilities bring great risks to cybersecurity. The National Information Security Vulnerability Sharing Platform (CNVD) included 14,201 security vulnerabilities in 2018, a decrease of 11% year on year. Among them, there were 4,898 high-risk vulnerabilities, a year-on-year decrease of 12.8%, but the highly threatening zero-day vulnerabilities increased by 39.6% year on year.[3] In the first half of 2019, 5,859 general security vulnerabilities were included onto the platform. Among them, there were 2,055 high-risk vulnerabilities.

6.2.1.2   The Frequency of DDoS Attacks Has Decreased, but Their Peak Traffic Continues to Increase

Since 2018, as China has carried out special campaigns to contain DDoS attacks, the frequency of DDoS attacks in China has generally shown a downward trend. According to the monitoring data of the National Internet Emergency Center (CNCERT), in the first half of 2019, the number of command and control (C&C) servers[4] used to launch DDoS attacks were 1,612. Among them, there were 144 or about 8.9% located in China, a decrease of 13% year on year. In contrast, the number of C&C servers located abroad doubled year on year. The total number of computers remotely controlled by hackers[5] was about 640,000, a year-on-year decrease of 10%, and there were about 6.17 million reflection attack servers, a year-on-year decrease of 33%. However, the peak traffic of DDoS attacks continues to increase. The growth in the number of attacks with domestic peak traffic above the TB/s level was greater than previous years. In 2019, there was an increasing number of cyberattacks which initiated a great number of access requests to target websites through public proxy servers. These attacks use fewer attack resources and bypass the CDN nodes configured by the websites, directly resulting in slow access to the websites and even the disruption of websites.

---

[3]"Zero-day" vulnerabilities are vulnerabilities against which no patch has been released when they are included by the CNVD.

[4]Through receiving commands from the command and control (C&C) servers, target machines will be controlled by C&C servers.

[5]The computers remotely controlled by hackers are controlled networking devices that receive commands from C&C servers and send out a large amount of traffic.

. 6.2.1.3 APT Attacks Against National Key Industrial Units Occur Frequently

APT attacks pose a serious threat to national security, economic development and civil rights and interests. According to the monitoring data of relevant institutions and public information and reports, government departments, state-owned enterprises, research institutes and energy, communication, military, nuclear and other infrastructure have been frequently attacked by advanced persistent threats.[6] Ocean Lotus, APT-C-09, Bitter, Darkhotel, Group 123, APT-C-01[7] and APT-C-12 are some public active APT organizations that are currently launching attacks in China. Ocean Lotus is one of the most active APT organizations that launch attacks in mainland China in recent years. It aims at a great number of targets in different fields, including government departments, large state-owned enterprises, financial institutions and research institutions. In the first half of 2019, its attack activities against many Chinese enterprises were detected repeatedly by relevant organizations. The APT-C-01 launches long-term cyberattacks against key units and departments such as national defense, government, science and technology, education and maritime institutions, and its attacks mainly focus on military industry, Sino-US relations, cross-strait relations, marine and other related fields. It drops vulnerability files or binary executable file through spear-phishing attacks. The APT-C-12 launches attacks through sending phishing emails and its attacks mainly focus on relevant information in nuclear industry, scientific research and other fields.

6.2.1.4 The Threat of Ransomware Over Critical Information Infrastructure in Important Industries Is Increasing

Since 2018, ransomware attacks have occurred frequently in China, the number of ransomware variants has been increasing, their influence and destructive power have been significantly enhanced, and their threat to cybersecurity has remained great. In 2018, the number of ransomware detected by CNCERT reached nearly 140,000, showing a growing trend on the whole. Among them, GandCrab, a ransomware with extensive attention, had 5 major versions and 19 mini versions in 2018. Another ransomware Lucky can launch rapid attacks by making use of weak password vulnerabilities and Window SMB vulnerabilities, which is thus difficult to prevent. Critical information infrastructure in key industries has gradually become the main target of the attack of ransomware. The government, healthcare, education and research institutions and the manufacturing industry are under grave ransomware attacks. For

---

[6]Data Source: *Global Advanced Persistent Threat Summary Report for 2018* released by QiAnXin Threat Intelligence Center, and *Advanced Persistent Threat Research Report for the First Half of 2019* released by Tencent Yujian Threat Intelligence Center.

[7]APT-C-01 is also called as "Qiongqi" or "Lvban" by other security manufacturers in China.

example, ransomware variants such as GlobeImposter and GandCrab have attacked many medical institutions in China and seriously impacted the operation of their information systems.

6.2.1.5 The Issue of Large-Scale Leakage of User Personal Information Remains Severe

Large-scale leakage of user personal information appears time and again, which continues to cause wide concerns and worries from all walks of life. In 2018, user personal information leakage incidents occurred frequently in China. More than 1 billion pieces of user information of express companies, 240 million pieces of check-in information of a chain hotel, and 9 million pieces of user information of a website were leaked, which included a large amount of personal privacy information such as user names, addresses, credit card numbers, ID card numbers, phone numbers and information of their family members. These incidents have brought heavy hidden danger to the personal safety of Internet users and property security. At the beginning of 2019, many widely-used databases in China, such as MongoDB and Elasticsearch, were found to have security vulnerabilities that might lead to data leakage. According to CNCERT's sampling and monitoring data, there were about 25,000 IP addresses used for MongoDB database services on Internet in China. Among them, the number of IP addresses with data leakage risk exceeded 3,000. In the 11,000 databases sampled, 43% had high-risk vulnerabilities and covered many important industries. Elasticsearch database was also found to have similar security risks. Among the 9,000 databases sampled, up to 73% had high-risk vulnerabilities. The above-mentioned two kinds of databases are mainly distributed in regions such as Beijing, Shanghai, Guangdong and Zhejiang, and are mainly owned by enterprises. These databases are very easily attacked. By default, local or remote databases can be accessed through default ports without permission verification, and any addition, deletion, modification, query and other operations can be performed in this way, thus resulting in high risk of data leakage.

6.2.2 New Cybersecurity Threats Continuing to Emerge

6.2.2.1 Fake and Counterfeit Mobile Applications Are Increasing and Becoming a New Channel of Internet Fraud

In recent years, with the deep penetration of Internet into economic and life fields, new ways of Internet-based remote non-contact fraud such as "online investment", "online dating" and "online shopping rebate" have continued to emerge. In 2018, the number of Internet fraud incidents based on mobile applications was particularly high. For example, many fake "loan apps" were used by fraudsters to defraud users of their privacy information and money. According to CNCERT's sampling and monitoring data, more than 1.5 million users have submitted their personal privacy information such as their names, ID card photos, personal asset certificates, bank account and addresses onto these fake "loan Apps", and many victims have suffered great loss of economic interests by paying "guarantee fees" and "service charges" of more than RMB ten thousand per capita to the fraudsters. In addition, the number of counterfeit apps that have similar icons or names with genuine software is rising. In 2018, CNCERT captured 838 new samples of counterfeit mobile Internet apps in the financial industry through means of independent monitoring and complaint reporting, with an increase of nearly 3.5 times year on year and achieving new high in recent years. These fake apps usually entice users to download and install them by means of "clout chasing", and may cause the leakage of personal privacy information such as the user's address list and SMS content. They may also download malicious software without the user's permission, resulting in malicious fee deduction and other harms.

Special Column: A Micro Loan Fraud Case—Loan Fraud by a Fake "XX Finance" App
The victim received a "loan promotion" call, asking he/she to add a customer service person on WeChat who sent the victim a QR code. After scanning the QR code, the victim jumped to the download page of a fake "XX Finance" app, which required the victim to fill in personal information to register on it. After successful login, the victim had to fill in personal sensitive information in order to apply for a "Loan". The operation diagram of micro loan fraud process is shown in Fig. 6.1.

Upon the completion of filling in relevant information, the victim was permitted to apply for a "Loan". After some period of time of application, an "Abnormal State" was shown for the loan order on the app, claiming that "Payment failure due to inconsistency between the bank

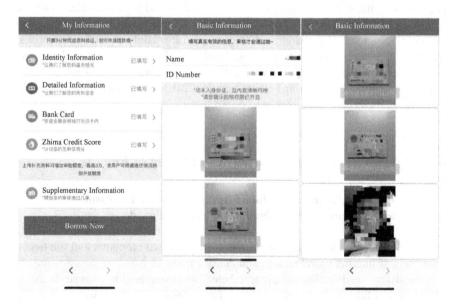

Fig. 6.1 Operation diagram of micro loan fraud process

account and account name. Due to your misoperation, the loan will not be granted and you will be on the list of dishonest persons". Then the victim received a "Customer Service" call, demanding him/her to pay a deposit equivalent to 20% of the loan amount through WeChat, Alipay or other channels, in order to avoid impacting personal credit. After payment, the victim neither received any loan nor had his/her commission charges and deposit back.

6.2.2.2 Cloud Platforms Become a Major Target of Various Cyberattacks

A large number of information systems are currently deployed on cloud platforms where massive data related to national economy and people's livelihood and business operations as well as personal information are an important target of cyberattacks. Besides, as cloud platform users pay less attention to cybersecurity protection, which further aggravates the cybersecurity risks. According to the monitoring data of relevant institutions, in 2019, the numbers of DDoS attacks in cloud platforms, links with backdoor programs and tampered web pages were 69.6%, 63.1% and 62.5% respectively of the total number of their categories in China. Meanwhile, as cloud services are characterized by great convenience and reliability, low cost, high bandwidth and high performance, and attackers can easily hide their real identities due to the complexity of cloud network traffic, many attackers launch cyberattacks by taking cloud platform devices as stepping stones or control ends. According to statistics, the number of DDoS attacks launched against the targets in China by utilizing cloud platforms accounts for 78.8% of the total number of DDoS attacks.

6.2.2.3 Networked Intelligent Devices Face Increasing Malicious Attacks

Despite the rising popularity of intelligent devices such as intelligent wearable devices, intelligent appliance and intelligent transportation devices in recent years, their capacity of security protection is generally weak and there exist such issues as weak passwords, inappropriate security configuration and imperfect upgrading and maintenance mechanism, indicating the existence of serious security risks. In 2018, 2,244 of the security vulnerabilities recorded by CNVD were related to networked smart devices such as home routers and web cameras, up 8% year on year. Examples of malicious programs that are currently active on networked smart devices include Ddosf, Dofloo, Gafgyt, MrBlack and Persirai. These programs and their variants have caused many hazards, including the leakage of user information and device data and the control over and destruction of hardware devices. The CNCERT's sampling and monitoring data suggested that in the first half of 2019, about 19,000 server IP addresses were controlled by malicious programs of networked smart devices, up 11.2% year on year, and about 2.42 million IP addresses of networked smart devices were controlled. Among them, there were nearly 900,000 IP addresses in China, a year-on-year decrease of 12.9%, and the daily average number of DDoS attacks launched by controlling networked smart devices was 2,118.

6.2.3 Key Industries Facing Serious Cybersecurity Risks

6.2.3.1 The Highly Risky Industrial Internet Is Increasingly Under Attack

Steady progress is currently made in the scale deployment and trial implementation of IPv6 and 5G in China, which will effectively promote the rapid development of industrial Internet as well as the grid expansion of critical information infrastructure in both width and depth. Meanwhile, however, the risks of cybersecurity and industrial security are intertwined, resulting in growing derivative threats. According to *China Internet Cybersecurity Report for the First Half of 2019*, the monitoring data of relevant institutions showed that there were a total of 6,814 exposed networked industrial devices in China, which were classified into 50 types such as programmable logic controllers, data acquisition and monitoring servers and serial servers, and were sourced from 37 well-known manufacturers at home and abroad such as Siemens, Vykon Controls and Rockwell. Among them, 34% were devices with high-risk vulnerabilities, and the manufacturers, models, versions, parameters and other information of these devices suffered malicious sniffing for a long time. The number of malicious sniffing incidents reached up to 51.51 million only in the first half of 2019. Besides, some large industrial cloud platforms in China have been under continuous cyberattacks, such as vulnerability exploitation, denial of service and brute force attack, which have caused great pressure on security protection.

6.2.3.2 Internet Financial Platforms and Mobile Applications Are Facing Severe Security Risks

In recent years, the operators of Internet financial platforms have raised their cybersecurity awareness and strengthened their cybersecurity protection capacity. However, some platforms are still short of cybersecurity protection capacity and facing many potential security problems. According to the monitoring data of relevant institutions, a great number of high-risk vulnerabilities were found in China's Internet financial platforms in the first half of 2019, including SQL injection vulnerability, remote code execution vulnerability and sensitive information disclosure vulnerability. In addition, CNCERT detected 505 security vulnerabilities in 105 Internet financial apps, and 239 were high-risk vulnerabilities.[8] Among these high-risk vulnerabilities, the number of plaintext data transmission vulnerabilities was the largest, up to 59,

---

[8] *China Internet Cybersecurity Report for the First Half of 2019* released by CNCERT, August 2019.

followed by plaintext password storage vulnerabilities for web views, the number of which was 58. The number of source code decompiling vulnerabilities was 40. These security vulnerabilities may threaten transaction authorization and data protection and cause data leakage. Some of them may impact the file protection of applications, making it impossible to effectively prevent the reversion or decompilation of applications, and have huge potential security risks.

6.2.3.3 Important Information Systems and Data in the Healthcare Industry Are Facing Severe Cybersecurity Challenges

In general, important information systems of the healthcare industry face relatively high security risks and multiple potential security problems. They are less capable of defending against public Internet attacks. The main challenges include:

(1) In the systems, there are high-risk vulnerabilities, severe issues of "botnets, Trojans, worms" and serious threats of ransomware.
(2) A large number of sensitive services are exposed; the weak password problem is outstanding; and data leakage incidents occur frequently.
(3) The versions of application components are low; there is high probability of website defacement; and illegal information is implicitly implanted in the systems.

According to the monitoring data of relevant institutions, 709 management systems for medical information, gene testing and other data in the healthcare industry were exposed on the public network in the first half of 2019, and 72% of them had high-risk vulnerabilities. Some of the monitoring or management systems that were exposed suffered from malicious sniffing and cyberattacks from abroad. Besides, China's biological data was transferred abroad for more than 1.42 million times, which involved more than 9,000 IP addresses and 658 units in China. Most of them was transferred to the United States, accounting for 33.4% of the total cross-border transfer times. China's medical imaging data was transferred abroad for more than 1.36 million times, which involved more than 1,400 IP addresses. The top three destinations were the United States, Vietnam and Canada, with the number of cross-border data transfer times reaching 90.56% of the total number.

6.2.3.4 Critical Information Infrastructure in Power Industry Becomes a Target of Attack at the National Level

Power blackout accidents caused by cyberattacks are common around the world in recent years. The power system is becoming a key target of national cyberattacks. CNCERT conducted a security test on the products of domestic major power manufacturers in the first half of 2019. It was found that under the guidance of power grid enterprises, power equipment suppliers became aware of the importance of cybersecurity, but the overall cybersecurity level of the power equipment still needed improving. As of June 2019, medium- and high-risk vulnerabilities had been detected more than 70 models of products in six categories, namely, measurement and control devices, protection devices, intelligent telecontrol equipment, station control software, phase measurement units, and cybersecurity situation-awareness and acquisition devices, from 28 manufacturers. These vulnerabilities might cause DDoS attacks, remote command execution, information leakage and other hazards. For example, the SISCO's MMS[9] Protocol Suite vulnerability may affect every power device that adopts the MMS protocol.

6.3 Positive Progress in the Cybersecurity Protection and Assurance Work

China has thoroughly enforced *Cybersecurity Law*, focused on improving the security protection level of critical information infrastructure, strictly implemented the system for classified protection of cybersecurity, and stepped up efforts in data security administration and personal information protection. Consequently, its cybersecurity protection capacity and level have been constantly improved, and its cybersecurity and people's lawful rights and interests have been effectively safeguarded.

6.3.1 Accelerating Efforts in the Security Protection of Critical Information Infrastructure

Critical information infrastructure is the top priority of cybersecurity protection. Chinese government has prioritized the security protection of critical information infrastructure in the cybersecurity work by strengthening job arrangements and implementing of security responsibilities. As the operators of critical information infrastructure, enterprises in finance, power, energy, transportation and other industries have actively shouldered the primary protection responsibilities. Relevant authorities have fulfilled their regulatory responsibilities by investigating risks and hazards, finding and solving hidden dangers and strengthening protection. With their efforts, the security protection level of critical information infrastructure has been improved effectively. In July 2017, *Regulations on Critical Information Infrastructure Security Protection (Draft for Comments)* was made public for open solicitation of opinions. It will provide important legal support for the security protection of critical information infrastructure. In order to advance the security testing

---

[9]The Manufacturing Message Specification (MMS) Protocol is a suite of communication protocols defined by the ISO Standard 9506 for industrial control systems. It aims to regulate the communication behaviors of the intelligent sensors, intelligent electronic equipment and intelligent control equipment that have communication capabilities in the industrial sector, and make system integration easy and convenient.

work of critical network devices and strengthen cybersecurity vulnerability management, in June 2019, the Ministry of Industry and Information Technology (MIIT) and relevant departments drafted *Implementing Measures on Security Inspection for Critical Network Equipment (Draft for Comments)* and *Administrative Provisions on Cybersecurity Vulnerability (Draft for Comments)* respectively for solicitation of public opinions. The National Internet Emergency Center (CNCERT) released *Standard System Framework of Critical Information Infrastructure Security Protection (Draft)*, which provided standard norms for critical information infrastructure security protection. In 2018, China's National Information Security Standardization Technical Committee carried out the pilot project for implementation of the standards in *Guidance for Security Inspection and Evaluation of Critical Information Infrastructure (Draft for Approval)*. Through trial implementation in 12 typical critical information infrastructure units from key industries and fields, the project verified the rationality and maneuverability of the standards and sought experience from the practices of security inspection and evaluation of critical information infrastructure according to relevant standards.

6.3.2   Upgrading and Improving the System for Classified Protection of Cybersecurity

The system for classified protection of computer information system security plays a critical role in the construction of the mechanism and capacity to protect cybersecurity. In June 2018, the Ministry of Public Security issued *Regulations on Classified Protection of Cybersecurity (Draft for Comment)*, indicating an upgrading from Classified Protection 1.0 (information security system protection) to Classified Protection 2.0 (cybersecurity protection). Classified Protection 2.0 has the following three characteristics:

(1) The basic requirements, evaluation requirements and design- technology requirements of classified protection are consistent in terms of their framework.
(2) General security requirements are combined with the requirements for security expansion of new applications, and cloud computing, mobile Internet, IoT and industrial control systems are included in standard specifications.
(3) Trusted verification is included as a main functional requirement at all levels and in all links.

In May 2019, the State Administration for Market Regulation and the Standardization Administration of the People's Republic of China officially issued a series of national standards related to Classified Protection 2.0, including *Information Security Technology—Baseline for Classified Protection of Cybersecurity*, *Information Security Technology—Evaluation Requirements for Classified Protection of Cybersecurity*, and *Information Security Technology—Technological Requirements of Security Design for Classified Cybersecurity Protection*, which were scheduled to be implemented on December 1, 2019. *Information Security Technology—Implementation*

Guide for Classified Protection of Information System, the national standard for Classified Protection 2.0, is also expected to be formulated and issued within the year.

6.3.3 Intensifying Data Security Administration and Personal Information Protection

China has strengthened data security administration, actively regulated the collection and use of personal information and effectively enhanced the level of data security administration and personal information protection through means such as constructing laws and regulations, carrying out special actions and formulating technological specifications.

(1) Perfect the legal system and provide important legal basis for data security and personal information protection. In 2019, the National Internet Information Office and relevant departments jointly researched and drafted a series of regulations, including *Measures for Data Security Management (Draft for Comments)*, *Measures for Cybersecurity Review (Draft for Comments)* and *Measures on Security Assessment of the Cross-border Transfer of Personal Information (Draft for Comments)*, for open solicitation of public opinions, aiming to consolidate the legal basis of data security administration and personal information protection. In August 2019, the National Internet Information Office issued *Provisions on the Cyber Protection of Children's Personal Information*. As the first legislation on children's cyber protection, it set strict rules on the collection, storage, use, transfer and disclosure of children's personal information.

(2) Reinforce the management and set severe penalties for those who harm personal information security according to the law and regulations. In 2019, the Office of the Central Cyberspace Affairs Commission, the Ministry of Industry and Information Technology (MIIT), the Ministry of Public Security and the State Administration for Market Regulation (SAMR) jointly carried out a special campaign against the collection and use of personal information by apps in violation of laws and regulations across the country. App operators were required to strictly fulfill their responsibilities and obligations under *Cybersecurity Law*, lawfully collect and use personal information, take responsibility for the security of personal information collected, and take effective measures to strengthen personal information protection. Since the launch of the special campaign, a great number of reports submitted by Internet users have accepted through the reporting channels. Evaluation agencies have evaluated the apps with a large number of users, and urged the rectification of those with serious problems through means such as sending rectification notices and making inquiries.

(3) Formulate technological specifications and effectively guide app operators to strengthen personal information protection. In order to address the current

pressing concerns of Internet users over some apps, such as compulsory authorization, claim of excessive authority and out-of-range collection of personal information, a series of documents were issued, including *Guide to the Self-Assessment of Illegal Collection and Use of Personal Information by Apps* by the working group on the special campaign against the collection and use of personal information by apps in violation of laws and regulations jointly established by the National Information Security Standardization Technical Committee, the Chinese Consumer Association, the Internet Society of China and the Cybersecurity Association of China, and *Cybersecurity Practices Guidelines—Specification of the Essential Information for the Basic Business Functions of Mobile Internet Apps* by National Information Security Standardization Technical Committee. They provide useful reference and guidance for app operators to conduct self-examination and self-correction, further regulate the order of cyberspace, and lay a foundation for establishing a long-term governance mechanism.

The above-mentioned efforts have caused great response in the society and played a positive role in guiding app operators to protect personal information.

6.3.4 Making Great Efforts to Improve the Security and Controllability Level of Cloud Computing Services

Cloud platform security is drawing increasing attention from all parties ranging from relevant state departments, cloud platform users and management operators. In 2014, *Opinions on Strengthening the Cybersecurity Management of the Cloud Computing Services of Party and Government Department* was issued, according to which cybersecurity reviews were launched on the cloud computing services of Party and government departments. In July 2019, the National Internet Information Office, the National Development and Reform Commission (NDRC), the Ministry of Industry and Information Technology (MIIT) and the Ministry of Finance jointly issued *Measures for Security Assessments of Cloud Computing Services*. It put forward to establish a systematic evaluation mechanism covering aspects such as evaluation subjects, responsibilities and processes, and provided support for improving the security and controllability level of the cloud computing services procured and used by Party and government organization and critical information infrastructure operators, as well as reducing the cybersecurity risks caused by the procurement and use of cloud computing services. As of August 2019, 14 government cloud computing service providers had passed evaluation.

6.4 More Efforts Made in the Handling and Special Rectification of Cybersecurity Incidents

China has effectively strengthened its response to cybersecurity incidents by carrying out various special campaigns against outstanding issues such as ransomware, DDoS attacks and illegal and criminal network activities, severely cracking down on law and regulation violations by law and maintaining a tough stance. In doing so, China cybersecurity protection and governance capabilities have been effectively improved.

6.4.1 Continued Efforts in the Handling of Cybersecurity Incidents

Relevant departments and institutions have effectively improved their capabilities of cybersecurity situation awareness and emergency response and properly handled a series of cybersecurity incidents. According to research reports, 48.5% of the ministries and commissions and 56.3% of the central enterprises in China have set up security operations centers. Their average response time of security incidents has decreased from about three days three years ago to less than one hour now.[10] In 2018, CNCERT successfully handled more than 100,000 cybersecurity incidents of various types, up 2.1% over 2017. Among them, the number of web spoofing incidents was the greatest, followed by security vulnerabilities, malicious programs, website distortion, webshell and DDoS attacks. In the first half of 2019, it coordinated and handled about 49,000 cybersecurity incidents, with a year-on-year decrease by 7.7%. In 2019, at the organization of CNCERT's industry alliances such as CCTGA,[11] CNVD and ANVA,[12] China's basic telecom operators, Internet enterprises, domain name registration management and service institutions, and mobile phone application stores carried out 14 special campaigns against public Internet malicious programs. As a result, they shut down 722 large botnets at home and abroad, and cut off hackers' attacks on nearly 3.898 million infected hosts, and withdrew 3,517 mobile Internet malicious Apps.

---

[10]Data Source: *2019 Research Report on the Development Trend of Cybersecurity Construction in Large and Medium-sized Government and Enterprise Institutions* released at 2019 Beijing Cybersecurity Conference.

[11]CCTGA, or the China Cyber Threat Governance Alliance, was established in 2016, with 90 enterprises first applying for its membership.

[12]ANVA, or the Anti Network-Virus Alliance of China, was established in 2009. It was jointly initiated by CNCERT/CC and basic Internet operators, cybersecurity manufacturers, value-added service providers, institutions on search engines and domain name registration and other units.

6.4.2   Solid Progress in the Management of Prominent Cybersecurity Issues

6.4.2.1   Special Campaigns Against Ransomware Are Launched

In early September 2018, the Ministry of Industry and Information Technology (MIIT) organized basic telecom operators, professional cybersecurity institutions, Internet enterprises and cybersecurity enterprises to jointly convene the Seminar on Special Campaigns against Malicious Programs to study the operation principle, transmission channels, prevention and disposal measures of ransomware. At the end of September, the MIIT issued *Notice on Carrying out Special Campaigns against Ransomware*, calling for joint monitoring and collaborative handling efforts of the communications administrations, telecoms and Internet enterprises and professional cybersecurity institutions across the country. The cybersecurity inspection results indicated that there were 60 security problems in the important systems of the basic telecom operators, including weak passwords and high-risk vulnerabilities.

6.4.2.2   Special Campaigns Against DDoS Attacks Are Launched

Since 2018, the CNCERT has focused on analyzing the network resources of DDoS attacks and carried out special campaigns to dispose of those resources. After one year of efforts, the monthly active number of domestic controlling devices, computers remotely controlled by "hackers" and other resources declined remarkably over 2017. According to external reports, China fell from the top three to tenth place in the world in terms of the number of domestic botnet controllers,[13] and the number of its DDoS active reflection sources decreased by 60%.[14]

6.4.2.3   Special Campaigns for Cybersecurity Are Carried Out

In order to help website operators to enhance their cybersecurity awareness and protection capacity and fulfill their cybersecurity obligations, from May to December 2019, the Office of the Central Cyberspace Affairs Commission, the Ministry of Industry and Information Technology (MIIT), the Ministry of Public Security and the State Administration for Market Regulation (SAMR) jointly carried out special campaigns to safeguard Internet cybersecurity across the country, including cleaning up the websites that had no archival filing or the archival filing information of which was inaccurate, severely cracking down on illegal and criminal activities, and imposing penalties on and exposing illegal websites. One of the distinctive features about this campaign was to tighten sanctions on the website operators that failed to

---

[13] Source: *DDoS Attacks in Q4 2018* released by Kaspersky Labs.
[14] Source: *2018 DDoS Attack Situation Report* jointly released by Telecom Yundi and NSFOCUS.

fulfill their cybersecurity obligations and caused cybersecurity incidents, and urge them to earnestly implement safety protection responsibilities and strengthen the management and maintenance of website security. During the campaign, the Office of the Central Cyberspace Affairs Commission strengthened overall planning and coordination and guided relevant departments to share information and collaborate with each other. Consequently, the cybersecurity awareness and protection capacity of website operators were effectively enhanced and website security was greatly improved.

6.4.3 Punishing Illegal Activities in Cyberspace by Law

The Ministry of Public Security launched "Clean Cyberspace 2018" with the public security departments across the nation in 2018, a 10-month special campaign to address eight cyber issues, including rampant hacker attacks, Internet enterprises' failure to earnestly fulfill their cybersecurity management obligations, and stealing of citizens' personal information for the purpose of trading. The public security departments across the nation have furthered the integration of management and control according to the overall arrangements of the Ministry of Public Security. Besides, they have always insisted on the combination of the essential work and special campaigns against illegal activities. On the one hand, they have promoted enterprises' strict implementation of the primary responsibility for security; on the other hand, they have resolutely cracked down on illegal and criminal network activities. As a result, more than 57,000 cybercrime cases were solved; more than 83,000 criminal suspects were arrested; more than 34,000 Internet enterprises and networking units were given administrative punishments; more than 35,000 apps with malicious programs or malicious behaviors were pulled from the shelves by law; and more than 20,000 illegal websites (columns) were shut down. Besides, the amount of key illegal information on online selling of guns, explosives, controlled knives, drugs and citizens' personal information decreased by about 30% over the previous year. Remarkable progress was also seen in the special campaigns against illegal activities.

6.5 Steady Progress in Cybersecurity Industry and Technology

The strong capacity to protect cybersecurity cannot be divorced from the powerful support of cybersecurity industry and technology. China has made concentrated efforts in developing cybersecurity industry and carried out active innovations in cybersecurity technology. A large number of cybersecurity enterprises have accelerated development, new products and services have continued to emerge, the comprehensive strength of cybersecurity industry has grown steadily, and the industrial scale has continued to increase. At the same time, China's technological support for the industry remains insufficient when compared with its ever-growing demand for cybersecurity of information infrastructure, service applications and data information, and the industrial scale remains to be further expanded.

6.5.1   Cybersecurity Industry Showing an Overall Good Momentum of Development

6.5.1.1   Enterprises Take on a Good Momentum of Development

In 2018, China had 17 enterprises listed among the world's top 100 in cybersecurity. Among them, Huawei ranked the highest, in the eighth place.[15] The business revenue has also maintained a steady growth trend on the whole. According to the financial reports of listed enterprises, the average business income of China's 10 most representative listed cybersecurity enterprises reached 1.569 billion *yuan* in 2018, up 10.69% over 1.418 billion *yuan* in 2017. In terms of net profit, the growth of corporate net profit has slowed down on the whole. The average net profit of the above-mentioned 10 enterprises was 268 million *yuan* in 2018, a slight increase of 6.67% over 2017, but much higher than the international growth rate of minus 53.14%. In terms of R&D investment, Chinese enterprises have intensified their efforts in research and development. The average R&D investment of the 10 enterprises was 267 million *yuan* in 2018, up 25.2% over 2017.

According to incomplete statistics, there were 2,898 enterprises engaged in cybersecurity in China in 2018. Beijing, Guangdong and Shanghai had the largest number, 975, 366 and 288 respectively. In 2018, the number of cybersecurity enterprises in China increased by 217.[16] See Table 6.1 for China's top ten regions with the largest number of cybersecurity enterprises.

6.5.1.2   Efforts Are Made in Promoting the Globalization of Relevant Products and Services

In recent years, a growing number of Chinese cybersecurity enterprises have formulated global development strategies and made great efforts to expand overseas cybersecurity market.

---

[15]Data Source: *2018 Global Cybersecurity Industry Investment and Financing Report* released by Shanghai Cybersecurity Industrial Innovation Research Institute.

[16]Data Source: Cybersecurity Industry Open Platform of the China Academy of Information and Communications Technology.

Table 6.1 China's top ten regions with the largest number of cybersecurity enterprises

Number	Region	Number of cybersecurity enterprises	Proportion (%)
1	Beijing	975	33.64
2	Guangdong	366	12.63
3	Shanghai	288	9.94
4	Jiangsu	167	5.76
5	Sichuan	139	4.80
6	Shandong	137	4.73
7	Zhejiang	120	4.14
8	Fujian	104	3.59
9	Hubei	82	2.83
10	Liaoning	70	2.42

(*Data Source* China Academy of Information and Communications Technology)

(1) Their overseas business income has achieved rapid growth. As of December 2018, the overseas business income of some Chinese enterprise grew by more than 70% over 2017.

(2) They have actively participated in international cybersecurity activities. A total of 36 Chinese enterprises participated in the RSA Information Security Conference held in March 2019 in the United States, an increase of 38% over 2017. On the conference, they presented a series of cybersecurity solutions, including cloud security, industrial control security, big data security, terminal security, identity management and access control, to global security vendors.

6.5.2 Emerging Technologies Boosting the Development of Cybersecurity Industry

6.5.2.1 5G Cybersecurity Receives Growing Attention

As new architectures and technologies such as network function virtualization, edge computing and network function opening are introduced into 5G network, fuzzy boundaries of device security, open ports, centralized controllers and edge deployment nodes in 5G network are constantly arousing new security demands. Most security enterprises in China are exploring ways to upgrade security products and services that suit 5G network characteristics and business features in preponderant domains. Security operators are mainly seeking security solutions for 5G network architecture and business scenarios. Security device manufacturers are committed to finding out and preventing security vulnerabilities of 0-day, 1-day and other devices. While developing secure 5G devices, they also work with security operators to work out security solutions for vertical industries.

6.5.2.2 The Integration of AI and Cybersecurity Is Accelerating

The integration pattern of "AI + cybersecurity" has gradually become the focus of China's current corporate research and innovation in the cybersecurity field:

(1) AI is utilized for defense. In recent years, in addition to traffic anomaly detection, malware detection, user abnormal behavior analysis and sensitive data identification, AI has also been deeply applied into areas such as cybersecurity situation awareness and threat intelligence mining and analysis.
(2) AI is utilized for AI security. In view of the new security threats caused by AI data samples, models/algorithms and implementation methods, as well as the resulting security issues such as data pollution, identification system disorder and software vulnerabilities, more attention has been paid to the study on the security of AI defense algorithms and platform frameworks.
(3) AI is utilized to respond to AI attacks. Due to the outstanding asymmetric effect of cyberspace attack and defense confrontation costs, AI may be utilized to reduce the attack cost and effectively conceal the identity of attackers. AI is currently gradually used in spear phishing, loophole attack and other areas. Besides, the concept of using AI to defend against AI attacks has also been gradually accepted and put into practice in AI industry.

6.5.2.3 "Zero Trust" Identity Management Is in the Initial Stage

As a new concept of identity management, Zero Trust breaks the previous security mindset of "safe inside" and "unsafe outside". In the new pattern, trust is first built up based on users' strongly-dependent information such as behaviors, attributes and context, and then differentiated and minimum access is granted to different identities. In this way, an appropriate level of trust as necessary for access is provided adaptively. With the opening of Zero Trust cybersecurity market, Chinese cybersecurity enterprises have deployed to some extent and developed technologies of multi-factor authentication, user and entity behavior analysis, software defined perimeter Micro-Segmentation.

6.5.3 Constantly Optimized Ecological Environment of Cybersecurity Industry

6.5.3.1 Multi-faceted Measures Are Taken to Boost the Healthy Development of Cybersecurity Industry

(1) Favorable cybersecurity policies are introduced constantly. In February 2019, the CPC Central Committee and the State Council issued *Outline of the Development Plan for the Guangdong-Hong Kong-Macao Greater Bay Area*, requiring to improve the cybersecurity in the Guangdong-Hong Kong-Macao Greater

Bay Area, establish and improve the information disclosure and early-warning mechanism for network and information security, intensify efforts in real-time monitoring, disclosure and early warning as well as emergency response, and build a comprehensive cybersecurity defense system. In March 2019, the State-owned Assets Supervision and Administration Commission (SASAC) issued *Measures for Assessment of the Operational Performance of Persons in Charge of Central Enterprises*, which included cybersecurity as an assessment index.

(2) A series of cybersecurity standards for different fields are introduced. In December 2018, the National Information Security Standardization Technical Committee officially released 27 national standards under its centralized management, which involved digital signature, electronic citizen identity, IoT, virus prevention, cyberattacks, password and other aspects.

6.5.3.2 Investment and Financing Activities Are Frequently Carried Out

Investment and financing activities are frequent in the cybersecurity field. China now has 135 cybersecurity enterprises having obtained financing and 100 enterprises having invested in cybersecurity. In 2018, the number of financing transactions in the cybersecurity field in China reached 79,[17] up 36.21% over 2017, and the total financing amount reached 7.21 billion *yuan*, an increase of 86.79% over 2017. Among them, 20 transactions had their financing amount exceeding 100 million *yuan*, 44 exceeding 10 million *yuan*, and 2 exceeding one million *yuan*. They mainly involved data security, cloud security, industrial Internet/IoT security and mobile security.

6.5.3.3 The Construction of National Cybersecurity Industrial Parks Accelerates

As huge open platforms, national cybersecurity industrial parks help improve the industrial chain of cybersecurity industry and generate the concentration effect. Since 2018, solid progress has been made in the construction of cybersecurity industrial parks in China.

(1) Beijing's national-level cybersecurity industrial park enters a substantial phase of construction. In January 2019, Beijing National Cybersecurity Industrial Park was officially established. The industrial park aims to become a domestic leading and world-class center for high-end, high-tech and high-value cybersecurity industries, and now has 10 cybersecurity enterprises including 360 Enterprise Security Group settled in.

(2) The first phase of Tianjin Binhai Information Security Industrial Park is about to be completed. The total investment of the industrial park is 4.5 billion *yuan*. The first phase is scheduled to complete in September 2019. The park now has

---

[17]Data Source: CCID Consulting, *2019 WhiteBook on China's Cybersecurity Development*.

gathered 4 important alliances, 6 national and provincial research centers and 13 core enterprises in China's information security industry.

(3) The construction of Wuhan National Cybersecurity Talent and Innovation Base has made progress. As of August 2019, 98 registered enterprises settled in the park, with a registered capital of 28.415 billion *yuan*, and the park contracted 63 projects with a total investment of 351.476 billion *yuan*. The top 20 cybersecurity enterprises in China have all settled in the base. A complete cybersecurity industrial chain is taking shape in Wuhan.

6.5.4 Industrial Organizations Driving the Development of Cybersecurity Industry

Industrial organizations play an active role in promoting the sustained and healthy development of cybersecurity industry.

(1) The Cybersecurity Association of China (CSAC) has focused on enhancing the self-discipline of cybersecurity industry by actively guiding various enterprises to fulfill their cybersecurity responsibilities, building a bridge for industrial communication and cooperation and promoting the healthy industrial development. In 2019, CSAC visited many domestic well-known cybersecurity enterprises and research institutions for in-depth investigation. It analyzed the current situation and problems of cybersecurity industry and put forward some countermeasures and suggestions to promote the industrial development from both capacity building and investment. It spearheaded the establishment a mechanism for regular negotiations on cybersecurity situation awareness, and launched campaigns to support the security protection of critical information infrastructure, including preparing general papers, establishing online training platforms and carrying out pilot training activities. Besides, as an industrial organization, it assisted in organizing the China Cybersecurity Week, Cybersecurity Exhibition of the World Internet Conference, China International Digital Economic Expo (CIDEE), the World 5G Conference and other events. It has also made efforts to create a favorable environment for the development of cybersecurity industry.

(2) The China Cybersecurity Industry Alliance (CCIA) has actively launched cybersecurity projects and activities. In 2019, the CCIA organized multiple meetings to study how to formulate standards such as *Mobile Apps Security Specification* and *Specifications and Conditions for Cybersecurity Product and Service Providers*. In September 2019, the CCIA held an exhibition of excellent cybersecurity solutions and innovative products, which aroused positive response in the industry. On the first exhibition held in 2018, the enterprises winning the award for new products with the most investing value received financial support of about 330 million *yuan* in total from investment institutions, which produced

great social effect. In addition, the CCIA has also actively led Chinese enterprises to "go global". For example, it organized enterprises to participate in RSA 2019, Trustech 2019, CYBER TECH and other events.

6.6 Consolidating the Basis of Cybersecurity Work

China has made solid progress in every aspect of the basic cybersecurity work by quickening the pace of cybersecurity personnel training and enhancing the protection awareness and skills of cybersecurity. As a result, the basis of China's cybersecurity work is further consolidated.

6.6.1 Intensifying Cybersecurity Personnel Training

The competition in cyberspace is essentially the competition of talents. A team of excellent cybersecurity talents is necessary to improve the cybersecurity level. In order to speed up the cybersecurity personnel training and explore new ideas, new systems and new mechanisms for personnel training, the Office of the Central Cyberspace Affairs Commission and the Ministry of Education launched a demonstration project for building first-class cybersecurity colleges. Xidian University, Southeast University, Beijing University of Aeronautics and Astronautics, Wuhan University, Sichuan University, the University of Science and Technology of China, and the Information Engineering University of the Strategic Support Force were the first to enter the demonstration project. Under the guidance and support of relevant departments, seven universities have deepened the construction of cybersecurity colleges. Significant progress has been made in expanding enrollment, innovating the training model, strengthening the cooperation between research institutes and enterprises and promoting the coordinated efforts of enterprises, universities, research institutions and consumers. In September 2019, the second batch of universities selected into the demonstration project for building first-class cybersecurity colleges were officially announced, indicating that China's first-class cybersecurity colleges would further play a demonstration and leading role. In February 2019, the General Office of the Ministry of Education issued *Work Priorities of Education Informatization and Cybersecurity in 2019*, which explicitly proposed to improve the support and guarantee ability of cybersecurity personnel, draft *Guide to the Core Courses of Cyberspace Security Postgraduates*, continue to strengthen the construction of cyberspace security, AI and other disciplines, accelerate the construction of new engineering and technological disciplines in the cyberspace field, promote university-industry collaborative education, guide and encourage qualified vocational colleges to provide cybersecurity-related specialties, and continue to expand the scale of cybersecurity-related personnel training. The China Internet Development Foundation (CIDF) has made full use of its cybersecurity fund to award excellent teachers and students in cybersecurity courses and support the training of cybersecurity talents. The China Association of Communication Enterprises has held the Industry Cybersecurity Skills Competition for seven consecutive years, with more 5,000 contestants participating in the competition every year. Since 2017, the association has been guiding the work of cybersecurity personnel competency certification and playing a positive role in cultivating cybersecurity talents.

6.6.2 Continuously Improving the Citizens' Protection Awareness and Skills of Cybersecurity

The biggest risk is not realizing any risks. Chinese government has made great efforts to improve the public's cybersecurity awareness and has held six consecutive China Cybersecurity Weeks. After six years of efforts, a favorable atmosphere in which everyone bears his share of the responsibility for cybersecurity and participates in the cybersecurity work has formed. The Sixth China Cybersecurity Week was held in September 2019, themed on "maintaining cybersecurity for the people and by the people". It was attended by a large number of enterprise leaders, well-known experts and academic leaders in the cybersecurity field from home and abroad, far more than those in the previous China Cybersecurity Week. A series of events were carried out during the China Cybersecurity Week. For example, the Cybersecurity Fair exhibited the latest achievements through interactive, experiential, informative and interesting means, and introduced many intelligent applications and interactive experience scenarios by the use of AI, VR and AR technologies. The 2019 International Anti-Virus Conference and similar forums comprehensively analyzed key issues such as cybersecurity industrial standards, data security and personal information protection. The "Cybersecurity at the Grassroots Level" publicity campaigns included a cybersecurity knowledge contest, which improved the public's cybersecurity protection capacity through entertainment.

Thanks to China's fundamental work in cybersecurity in recent years, such as publicity, education and training campaigns, the security awareness and security sense of Internet users have been significantly improved. According to *2018 Survey Report on Internet Users' Sense of and Satisfaction with Cybersecurity* released on the 2018 Internet Security and Governance Forum,[18] Chinese Internet users' sense of and satisfaction with cybersecurity are in medium high level. More than 80% of Internet users believe that the security of the networks that they use in everyday life is above average, and nearly 40% believe it safe and very safe to use Internet. Chinese Internet users hold a positive attitude to strengthening cybersecurity governance. Nearly 90% of them believe it necessary or very necessary to strengthen cybersecurity governance, among whom 62.29% of the respondents believe it very

---

[18]Data Source: *2018 Survey Report on Internet Users' Sense of and Satisfaction with Cybersecurity* released on the 2018 Internet Security and Governance Forum.

necessary, and 24.55% believe it necessary. Internet users are actively and conscientiously participating in cybersecurity governance. Most of them make a judgment before responding to illegal Internet information. More than 40% verify the authenticity by searching relevant information, more than 30% make the judgment based on their own experience, and half of them seek help from relevant departments or units. According to the 44th "Statistical Report on China Internet Development" released by the China Internet Network Information Center (CNNIC),[19] there were more Internet users who did not encounter any cybersecurity problems when they used Internet. The rate was 55.6% in the first half of 2019, up 6.4% compared with the end of 2018. The foundation of "maintaining cybersecurity for the people and by the people" is increasingly consolidated.

On the whole, China has made solid progress and achieved positive results in all the cybersecurity work, and its level and capacity to protect cybersecurity have been significantly improved. However, as various cybersecurity risks intertwine with each other and are highly interconnected in China, the cybersecurity situation is becoming increasingly serious and complicated, and the future cybersecurity threats and risks will continue aggravating. In order to strengthen China's capabilities to uphold national cybersecurity and effectively safeguard national security and interests in cyberspace, it is necessary to further improve the cybersecurity guarantee system, strengthen the security protection of critical information infrastructure, and intensify data security administration and personal information protection. Other moves include actively developing the cybersecurity industry and further advancing the popularization of cybersecurity knowledge and skills.

---

[19]The 44th "Statistical Report on China Internet Development" released by China Internet Network Information Center (CNNIC), August 30, 2019, see http://www.cnnic.net.cn/hlwfzyj/hlwxzbg/hlw tjbg/201908/t20190830_70800.htm.

Chapter 7
Rule of Law Construction in Cyberspace

7.1 Outline

Internet development has not only spawned a series of new social relations, but also brought forth new social contradictions and conflicts. Emerging issues such as diffusion of harmful information, personal privacy disclosure and illegal data collection pose new challenges to cyberspace governance. General Secretary Xi Jinping pointed out that cyberspace was not a place beyond the rule of law, and cyberspace must be governed, operated and used in accordance with law, so that Internet could enjoy sound development under the rule of law.

Over the past year, the rule of law has played an increasingly fundamental role in cyberspace governance and the rule of law construction in cyberspace has been constantly improved. Specifically, the legal system in cyberspace affairs has been further improved. *E-Commerce Law of the People's Republic of China* ("*E-Commerce Law*") was officially promulgated and implemented. The top-level design of cybersecurity laws represented by *Cybersecurity Law of the People's Republic of China* ("*Cybersecurity Law*") has been basically completed, and relevant supporting laws and regulations have also been released successively. Besides, positive progress has been made in introducing important legislations such as *Personal Information Protection Law of the People's Republic of China* ("*Personal Information Protection Law*"), *Data Security Law of the People's Republic of China* ("*Data Security Law*") and *Cryptography Law of the People's Republic of China* ("*Cryptography Law*"), with legal framework continuously improved. China has also strengthened its enforcement of Internet laws. Relevant departments have continued to carry out security inspection of network facilities, cleanse illegal and harmful information, crack down on illegal behaviors such as violating citizens' personal information and spreading Internet rumors, and resolutely safeguard the order in cyberspace according to *Cybersecurity Law*, *E-Commerce Law* and other laws and regulations. The reform of developing a networked, transparent and intelligent judicial system is further deepened. The special adjudication mechanism represented by Internet courts has gradually matured. With the deep integration of institutions and technologies, the adjudication rules of Internet cases are gradually improved and relevant cases are handled in a more efficient and convenient manner. The depth and breadth of Internet law popularization and education are further expanded and the public's legal sense of cybersecurity is raised significantly.

7.2   Accelerated Introduction of Internet-related Laws and Constant Improvement in the Cyberspace Legal System

"The law is of great value in the governance of a country, and good laws are a prerequisite for good governance." Speeding up the development of Internet legislations is an important step to build the comprehensive Internet governance system proposed at the 19th National Congress of the CPC. With the rapid development of China's legal system on cyberspace in recent years, fruitful legislative achievements have been made in cybersecurity maintenance, network information services, Internet-based social management and other areas. On the one hand, a series of special Internet legislations on new applications (e.g., Internet-based e-commerce applications) and new technologies (e.g., blockchain) such as *E-Commerce Law* and *Provisions on the Administration of Blockchain Information Services* have been introduced. On the other hand, the top-level design of cybersecurity legislations represented by *Cybersecurity Law* has been basically completed. Key legislations such as *Regulations of the People's Republic of China on Telecommunications*, *Personal Information Protection Law*, *Data Security Law* and *Cryptography Law* have been listed in the legislative plan of the Standing Committee of the 13th National People's Congress, and the legislation progress is accelerating. Besides, relevant departments are actively formulating supporting laws and regulations on data governance, cross-border data flow, etc. At the same time, in order to address the new challenges and new problems brought by Internet development to traditional sectors, traditional legislations such as *Criminal Law of the People's Republic of China*, *General Rules of the Civil Law of the People's Republic of China*, *Anti-Unfair Competition Law of the People's Republic of China* ("*Anti-Unfair Competition Law*") and *Advertising Law of the People's Republic of China* ("*Advertising Law*") have also made timely adjustments and revisions.

7.2.1   Improving Supporting Laws and Regulations on Cybersecurity

Over the past year, relevant departments have further improved the supporting laws and regulations of *Cybersecurity Law* in accordance with the new situation, challenge and threat in cybersecurity protection.

7.2.1.1 The Requirements concerning the Security of Critical Information Infrastructure and Network Products and Services are Put into Practice

Cybersecurity Law makes special provisions on the responsibilities and obligations of relevant parties on the security protection of critical information infrastructure at the state, industry and operator levels. The implementation of *Regulations on Critical Information Infrastructure Security Protection* has been included in *2019 Legislative Work Plan of the State Council*, which is now being accelerated. In May 2019, the National Internet Information Office issued *Cybersecurity Review Measures (Draft for Comments)*. In June and July 2019, the Ministry of Industry and Information Technology (MIIT) solicited public opinions on *Implementing Measures on Security Inspection for Critical Network Equipment (Draft for Comments)* and *Administrative Provisions on Cybersecurity Vulnerability (Draft for Comments)*, and approved relevant projects for implementing the compulsory national standard *General Security Technical Requirements for Specialized Cybersecurity Products*. In doing so, they aimed to implement relevant requirements in *Cybersecurity Law*, such as establishing the cybersecurity review mechanism for the operators of the network critical information infrastructure to purchase cyber products and services, improve the security and reliability level of critical information infrastructure and strength cybersecurity management.

7.2.1.2 The Security and Reliability Level of Cloud Computing Services are Enhanced

With the gradual application and popularization of cloud computing services, higher requirements are raised for the stability and controllability of relevant facilities and networks. Among them, "Government Cloud" and large "Enterprise Cloud" platforms host a huge amount of data on the country, industries and people's livelihood, and their security has thus become the top priority. According to *Opinions on Promoting Innovation and Development of Cloud Computing and Cultivating New Formats for the Information Services Industry* issued by the State Council, cloud computing will become an important form of China's informatization and an important pillar of building China's strength in cyberspace by 2020. In July 2019, the National Internet Information Office, the National Development and Reform Commission (NDRC), the Ministry of Industry and Information Technology (MIIT) and the Ministry of Finance jointly issued *Measures for Security Assessments of Cloud Computing Services*, in a bid to further improve the security and reliability level of the cloud computing services procured and used by Party and government organization and critical information infrastructure operators.

7.2.1.3   The Life-cycle Security Management for Data is Strengthened

In May and June 2019, the National Internet Information Office solicited public opinions on *Measures of Data Security Management (Draft for Comments)* and *Measures on Security Assessment of the Cross-border Transfer of Personal Information (Draft for Comments)*. Under the framework of *Cybersecurity Law*, it further clarified and refined the security rules of data collection, processing, use and cross-border transfer. Meanwhile, local governments have constantly raised their data security awareness and actively formulated relevant local laws. In August 2018, *Measures for Big Data Security Management in Guiyang City*, the country's first local regulation on big data security management, was officially issued, which made clear provisions on the security responsibilities of relevant subjects in big data development and application. In June 2019, *Tianjin's Interim Measures for Big Data Security Management* started pilot implementation, which aimed to strengthen the overall planning and coordination of local data security work and set up a sound local data security assurance system.

7.2.2   Regulating and Strengthening Internet Information Services

The report of the 19th National Congress of the CPC proposed to enhance the construction of online content to ensure a clean cyberspace. However, due to the real-time, wide-range, interactive and virtual features of Internet, an increasing amount of information spreads faster in a wider range and the supervision of information dissemination becomes more difficult, which pose new challenges to web content management.

7.2.2.1   The Structure of Online Content Governance is Enriched

As Internet-based applications and their content become more diversified in recent years, competent authorities have issued specialized content management regulations on new information dissemination media such as instant messaging, search engines, apps, livestreaming, forum communities, groups and microblogs according to the principle of "comprehensive governance". Over the past year, relevant authorities have further improved the management framework for Internet information services and regulated the information services with public opinion attributes and blockchain information services. In November 2018, the National Internet Information Office and the Ministry of Public Security jointly formulated and issued *Provisions on the Security Assessment of Internet Information Services with Public Opinion Attributes or Social Mobilization Capabilities*, in order to urge and guide relevant information service providers to fulfill their security management obligations prescribed by law and prevent the harm caused by illegal information such as rumors and false information. In January 2019, the National Internet Information Office issued *Provisions on the Administration of Blockchain Information Services*, which clearly stipulated that archival filing was compulsory for blockchain information services and that the service providers should perform their security obligations, including real identity authentication and the formulation and disclosure of management rules and platform conventions.

7.2.2.2 The Management of Internet Financial Information is Strengthened

Financial information not only helps drive the development of financial products and market innovation, but also may mislead the market and heighten market volatility. Despite the booming in China's financial information service industry, the application form and content requires supervision in recent years. In February 2019, *Provisions on the Management of Financial Information Services* formulated by the National Internet Information Office was formally enforced, the purpose of which was to improve the quality of financial information services and promote the healthy and orderly development of financial information services. One the one hand, it required financial information service providers to raise their self-discipline consciousness and carry out financial information content management. On the other hand, it clearly stipulated that financial information service providers should obtain relevant qualifications for engaging in Internet news information services, legal franchise businesses or financial businesses that should be filed, and accept the supervision and management by relevant competent authorities.

7.2.3 Constantly Improving Laws and Regulations on Internet-based Social Management

In the past year, Internet has played an increasingly prominent role in safeguarding and improving people's livelihood and strengthening and innovating social governance. Relevant laws and regulations have been introduced constantly. Relevant legislations focus on adjusting the emerging Internet legal relationships, safeguarding the rights and interests of specific subjects and improving the legal means of Internet governance.

7.2.3.1 The Chaos in the E-commerce Field are Rectified

The flexible application of Internet has spawned new competition and profit-making patterns, calling for the formulation of special legislations to adjust new social relations stretching from these special attributes and address the new impacts on and new problems in traditional social relations. In recent years, the rapid development of e-commerce industry has created new business models and spawned new legal relationships involving merchants, platforms and consumers. In particular, the distribution of the platforms' management obligations and merchants' responsibilities has received wide attention. *E-Commerce Law*, enforced in January 2019, explicitly stipulates the rights and obligations of all parties involved in e-commerce activities. Besides, it strengthens the protection of consumers' rights and the restrictions on improper business and selling behaviors.

As Internet is an extension of the real world, many existing problems in the real world start to be reflected in the online world, the tackling of which requires timely revision of existing laws and regulations. Internet marketing has emerged with the birth and development of e-commerce. Due to the special attributes of Internet such as virtuality, improper online marketing behaviors occur frequently despite repeated prohibitions. According to *Advertising Law* revised in October 2018, the use of Internet to engage in advertising activities should not affect the normal use of Internet, and there should a close mark conspicuously indicated to ensure one-click close. In April 2019, the State Administration for Market Regulation (SAMR) solicited public opinions on the revised *Measures for the Supervision and Administration of Network Transactions*. The public opinions suggested that network transaction operators should provide non-individualized options and respect and protect the legitimate rights and interests of consumers while carrying out "precision marketing" activities based on their personal characteristics.

7.2.3.2 The Protection of the Rights and Interests of Minors and Other Special Subjects is Strengthened

Minors are more vulnerable to the interference and influence of Internet than adults due to the immaturity of their cognitive and self-control capabilities. In the past year, relevant departments have further strengthened the protection of minors' rights and interests on the Internet, improved special regulations on the protection of children's personal information and revised *Law of the People's Republic of China on the Protection of Minors* and *Regulations on the Protection of Minors Online*. In August 2019, the National Internet Information Office issued *Provisions on the Cyber Protection of Children's Personal Information* to specify the protection requirements of children's personal information on such aspects as the settings of children's special user agreement, the management responsibilities of internal management specialists, the consent of children's guardians, encrypted storage and minimum authorized access.

7.2.3.3 Network Social Governance Means are Improved

Network credit management has further improved and enriched the means of network social governance. In *Cybersecurity Law*, it is clearly stipulated that illegal acts shall be recorded in credit archives and disclosed to the public. At the beginning of 2018, the Ministry of Industry and Information Technology (MIIT) and the Ministry of Civil Affairs issued separate documents to introduce the list of bad business operations and the list of bad credit into the operation and management of telecommunication services and the management of social organization activities. In July 2019, the National Internet Information Office solicited public opinions on *Measures on Credit Administration for Seriously Untrustworthy Internet Information Services Entities (Draft for Comments)*, and planned to implement credit blacklist management and joint punishment for dishonest behaviors. In August 2019, the National Development and Reform Commission (NDRC) and relevant departments jointly formulated and sought public comments on *Implementing Opinions on Administration of Lists of Discredited Parties Subject to Joint Punishment in the Transportation and Logistics Sectors (Draft for Comments)*. According to the document, if the market entities and relevant persons in the transportation and logistics sectors infringe upon personal information or conduct other illegal acts, they may be identified as seriously dishonest persons and be blacklisted.

7.3 Intensified Efforts in Internet Law Enforcement to Facilitate the Standard and Orderly Development of Cyberspace

The vitality and authority of law both depend on its implementation. Over the past year, Internet law enforcement activities in areas such as personal information protection, network information content management, cybersecurity protection and the monitoring of network products and services have been carried out more frequently, more vigorously and more widely. The National Internet Information Office, the Ministry of Industry and Information Technology (MIIT), the Ministry of Public Security, other industry authorities and local government agencies have further enhanced the law enforcement capacity, refined their law enforcement measures and gradually improved relevant working mechanisms. They have actively implemented legislative requirements, cracked down on illegal acts and safeguarded market order and users' rights and interests through regular or random supervision and inspection.

7.3.1   Continuously Carrying out Cybersecurity Campaigns

In order to improve the construction, management, precaution and reform in cyberspace, relevant departments in China have strengthened the cybersecurity inspection work. With the clear knowledge of existing risks, they have investigated loopholes and hidden dangers, disclosed inspection results and urged relevant units to rectify their problems, thus having comprehensively improved their cybersecurity management and assurance level.

7.3.1.1   Network and Device Security

The stable and reliable operation of network and devices depends on measures to prevent cyberattacks, intrusions, interference, destruction, illegal use and accidents. In April 2019, the first national campaign to inspect the cybersecurity of critical information infrastructure was launched, the purpose of which was to identify the information systems and industrial control systems that would affect the operation of key businesses, grasp the security situation of China's critical information infrastructure, and provide basic data and reference for the construction of the security assurance system for critical information infrastructure. In June 2019, the Ministry of Industry and Information Technology (MIIT) carried out administrative inspection on the cybersecurity of telecom and Internet industries, the scope of which focused on anti-attack, anti-intrusion, anti-tamper and anti-stealing measures. In the campaign, the MIIT worked hard to find hidden cybersecurity risks, strengthened rectification measures and urged basic telecom operators, domain name registration management and service institutions and Internet service providers to implement their primary responsibilities. Besides, it strengthened the construction of cybersecurity protection capacity and guarded against major cybersecurity risks to ensure the continuous stable operation and data security of the telecommunication network and public Internet.

7.3.1.2   Information and Data Security

In order to ensure the integrity, confidentiality and availability of information and data, efforts shall be made to protect not only the storage security of network information, but also the security of information during its production, transmission and use. In July 2019, the Ministry of Industry and Information Technology (MIIT) issued *Special Action Plan for Improving the Network Data Security Protection Capability of the Telecom and Internet Industries*, according to which it carried out data security inspection on all the basic telecom operators, 50 key Internet enterprises and 200 mainstream mobile applications (App) in China and urged them to further improve network data security systems and standards. Other measures included carrying out data security evaluation, strengthening the data security management policies and procedures, promoting the construction, supervision and publicity of data security capability and improving the capacity to protect network data security.

7.3.2   Cleaning Up and Rectifying Illegal and Harmful Internet Information

Illegal and harmful information such as obscene, pornographic, violence and terrorism is a malignant tumor in cyberspace as it not only pollutes the network ecosystem and destroys the information dissemination order, but also seriously endangers the physical and mental health of teenagers. It is also deeply loathed by the public. In the past year, the National Internet Information Office, the Ministry of Industry and Information Technology (MIIT), the Ministry of Public Security and relevant departments have intensified law enforcement to rectify illegal and harmful information on Internet and guide enterprises to strengthen self-discipline and implement their primary responsibilities. Besides, sustained heightened efforts have also been made in rectifying illegal and harmful information.

7.3.2.1   Cleaning Up Vulgar and Harmful Information

In the first half of 2019, the National Internet Information Office launched a special campaign of network ecosystem governance to clean up 12 types of negative and harmful information (namely, pornographic information, vulgar information, violent information, horrifying information, gambling and fraud information, Internet rumors, feudalistic superstition information, abusing and spoofing information, threatening information, clickbait information, hate mongering information, and the information that spreads unhealthy lifestyles and unhealthy popular culture) on various websites, mobile clients, forums and post bars, instant communication tools, livestreaming platforms and other important links. As of June 12, 2019, more than 110 million pieces of obscene, pornographic, gambling and fraud and other harmful information had been cleaned up, over 1.18 million illegal accounts that spread pornographic and vulgar information, false rumors and other information in various platforms had been removed, and 4,644 registered websites had been closed or canceled. Since March 2019, the Office of the National Work Group for "Combating Pornography and Illegal Publications" has launched a series of special campaigns such as "Clean Cyberspace 2019", "Protecting Juveniles on the Internet 2019" and "Autumn Wind 2019" to continuously purify the social and cultural environment.

7.3.2.2 Combating Internet Water Army and Cyber Violence

At the beginning of 2018, the Ministry of Public Security organized a special campaign to investigate and combat Internet water army's illegal and criminal activities and crack down on the Internet water army groups that carried out illegal and criminal activities on Internet in the names of "public opinion supervision", "legal supervision" and "social supervision". According to statistics, in 2018, more than 50 law-breaking and crime cases of Internet water army were solved, more than 200 suspects were arrested, more than 20,000 related websites were shut down and more than 1,000 Sina V accounts were closed in the campaign. Besides, the Internet information offices and public security departments in all localities have also carried out timely investigations and made penalty decisions on the problems of Internet information distribution platform users spreading rumors that insulted and libeled others, infringing on their reputation, privacy and other legitimate rights and interests.

7.3.2.3 Management of Online Extortion and Paid Post Deletions

In the first half of 2019, the National Internet Information Office launched a campaign against online extortion and paid post deletions, through which it shut down nearly 300 illegal websites, closed more than 1.15 million illegal social network accounts, cleaned up more than 9 million pieces of illegal and unhealthy information, held talks with 136 websites, and closed down nearly 50 local and professional channels of central news websites. In addition, the Ministry of Industry and Information Technology (MIIT) imposed penalties on 80 illegal websites involving online extortion and paid post deletions. During the campaign, the State Administration of Press, Publication, Radio, Film and Television (SARFT) investigated almost 100 cases on fake media, fake correspondent station and fake reporters.

7.3.3 Intensifying Law Enforcement on Internet Society Management

Over the past year, relevant departments have strengthened information sharing and collaboration and repeatedly carried out joint law enforcement campaigns to crack down on relevant high-risk illegal acts, and a long-term supervision mechanism has gradually been established.

7.3.3.1 Cracking Down on Illegal Trading of Personal Information

National and local cyberspace affairs, public security, industry and information technology departments and other authorities have carried out actions to crack down on data violations. Take local law enforcement as an example. In the special campaign "Clean and Safe Cyberspace" launched by the Guangdong Department of Public Security, a total of more than 5,000 cyber crime cases were solved, more than 21,000 suspected were arrested and more than 730 million pieces of citizens' personal information were seized. In July 2018, Shandong uncovered an extraordinarily serious case of violation of citizens' personal information. As a result, 57 suspects were arrested, 11 companies involved were shut down and billions of pieces of personal information of citizens (up to 4,000 GB) were seized. These companies transmitted more than 130 million pieces of citizen's personal information every day.

7.3.3.2   Carrying out Evaluation on User Privacy Policies

Although the extensive development and application of apps have brought a lot of convenience to the public in recent years, they are also suffering from issues of excessive collection and infringement of user information. From August to October 2018, the Chinese Consumer Association carried out evaluation on apps' personal information protection and issued a report accordingly to warn against the risks of identical texts, imparity terms, standard terms and general authorization. At the beginning of 2019, the Office of the Central Cyberspace Affairs Commission, the Ministry of Industry and Information Technology (MIIT), the Ministry of Public Security and the State Administration for Market Regulation (SAMR) jointly issued *Announcement on Conducting the Special Campaign against the Collection and Use of Personal Information by Apps in Violation of Laws and Regulations*, according to which a special campaign against the collection and use of personal information by apps in violation of laws and regulations was carried out nationwide. Besides, some major apps were evaluated and the review results were announced in July to urge some law-breaking enterprises to make timely rectifications.

7.3.3.3   Regulating the Internet Market Order

With the rapid development of e-commerce, a series of illegal business activities such as Internet-based price fraud, click farming, illegal promotion and illegal tied sale have sprung up endlessly. In June 2019, the State Administration for Market Regulation (SAMR), the National Development and Reform Commission (NDRC), the Ministry of Industry and Information Technology (MIIT), the Ministry of Public Security, the Ministry of Commerce, the General Administration of Customs (GAC), the National Internet Information Office and the State Post Bureau jointly announced their plan to carry out the 2019 Online Market Regulation Special Action. According to *E-commerce Law, Anti-Unfair Competition Law* and other laws and regulations, they have regulated the archival filing of e-commerce entities, severely cracked down on outstanding market problems such as online sales of counterfeit and shoddy products, unsafe food, fake medicines and medicines of inferior quality, strengthened the monitoring of online transaction information and the random checking of product quality, supervised the e-commerce operator's fulfilment of responsibility, and investigated strictly improper acts of e-commerce platforms such as "choosing one from two".

7.3.3.4 Cracking Down on Illegal Internet Advertising

At the beginning of 2019, the State Administration for Market Regulation (SAMR) organized a special campaign against illegal Internet advertising. All the local industry and commerce departments and market regulation departments have actively deployed and carried out relevant work and focused on investigating key and major cases. Besides, member units of joint conferences have held talks with and inspected relevant companies to crack down on fake and illegal Internet advertising. According to statistics, in 2018, 23,102 illegal Internet advertising cases were handled by the market regulation authorities in China, accounting for 55.9% of the total number of advertising cases.

7.3.3.5 Cracking Down on the Piracy of Online Products and Games

Network copyright protection is vital to the development of Internet creative industry. In recent years, China has intensified its efforts in network copyright protection and carried out the special Cyber Sword Special Campaign every year. In April 2019, the National Copyright Administration of China launched the "Cyber Sword 2019" action jointly with the National Internet Information Office, the Ministry of Industry and Information Technology (MIIT) and the Ministry of Public Security to address the network infringement problems that drew sharp reactions from enterprises and the masses. According to statistics, from 2013 to 2018, China's copyright law enforcement departments at all levels handled 22,568 infringement and piracy cases, including cyber cases, and shut down 3,908 infringement and piracy websites by law.

7.4 Internet Justice Playing a More Important Role with the Trial Mechanism Improved Gradually

Facing the new characteristics, new rules and new demands brought by new Internet technologies and new industrial revolutions, China's justice model is in urgent need of renovation. In recent years, the reforms of developing a networked, transparent and intelligent judicial system such as network justice and intelligent court, have gradually evolved from a management solution to an important direction of judicial reforms. Over the past year, Internet courts have made satisfactory achievements in their reform and demonstration projects. As grassroots people's courts that have centralized jurisdiction over Internet cases, Internet courts adopt the new trial mechanism of "online trial of online cases". They have summarized replicable and general experience in fields such as case trial, platform construction, litigation rules, technological applications and Internet governance, improved the litigation results, and enabled the public to better understand the fairness, convenience and high efficiency of justice.

7.4.1 Significant Achievements in Internet Court Construction

Since the establishment of Internet courts two years ago, the trial mechanism and efficiency have witnessed steady improvement. As of June 30, 2019, Hangzhou Internet Court, Beijing Internet Court and Guangzhou Internet Court had handled 72,829 Internet cases and concluded 47,073 cases. The rate of application for online case trial was 94.72% and 35,267 cases were concluded online throughout the whole process. Online court hearing took an average of 45 min, saving about 60% compared with the traditional trial model. The average period of case trial was 38 days, about 50% less than that of the traditional trial model. The rate of willingly accepting first instance judgments reached up to 98%, and the trial quality, efficiency and result showed a good momentum of growth.

In order to improve the trial efficiency, Internet courts have insisted on adopting the mechanism for non-judicial dispute settlement, promoted one-stop multi-factor dispute resolution on a platform with full online process, full coverage of disputes and full online and offline integration, and provided appropriate personnel and plan to help with dispute settlement. For example, Hangzhou Internet Court has helped e-commerce platforms such as Taobao to establish a pre-litigative mechanism for settlement of complaints and disputes and give full play to their role of independent dispute settlement, thus helping resolve disputes from the source. The multi-factor dispute resolution platform of Guangzhou Internet Court brings together 25 mediation agencies and 284 mediators in the Guangdong-Hong Kong-Macao Greater Bay Area. Among them, there are more than 20 professional mediators from Hong Kong and Macao who can effectively resolve disputes involving Hong Kong and Macao.

7.4.2 Promoting Deep Institution-Technology Integration

Internet Courts actively respond to people's judicial needs in the new era, give full play to policy advantages and technological strength, and vigorously build an inclusive, equal, convenient, efficient, intelligent and accurate online litigation service system. Through the Internet litigation platform and the "Mobile Micro Court" Applet, they provide "one-stop" online litigation services and bring the whole trial online. The parties involved can participate in all litigation activities without leaving their homes, realizing "justice without trips". At present, of all the lawsuits tried at Beijing Internet Court, 100% of them were filed online, 90.29% of the litigation fees were paid online, and 96.82% of the judicative papers were delivered online.

In addition, Internet Courts are actively exploring the "blockchain + justice" model and building a judicial big data platform based on big data, cloud storage and blockchain technology, to realize data sharing and interoperability, research and identify litigation trends, and support social governance. They are also expanding the application of blockchain in trial and execution to solve the problems with evidence collection, preservation and authentication in litigation. In September 2018, the Supreme People's Court issued *Regulations of the Supreme People's Court on Several Issues concerning the Trial of Cases by Internet Courts*, confirming the legal effect of blockchain deposit as proof in Internet cases. In September 2018, Hangzhou Internet Court launched a judicial blockchain system, making it the first court in China to apply blockchain technology into mediate and settle disputes. In December 2018, Beijing Internet Court launched "balance chain", a judicial blockchain system, which recorded every digital data transmission with whole-chain credibility and all-node verification. In March 2019, Guangzhou Internet Court launched the "Netcom Legal Chain" electronic evidence system. Based on the underlying blockchain technology, the court built a smart credit ecosystem consisting of "one chain and two platforms—the Netcom Legal Chain, a creditable electronic evidence platform, and a judicial credit co-governance platform". Within a week of trial operation, the system registered over 260,000 data deposits.

7.4.3 Exploring Adjudication Rules for Internet Cases

In the past year, Internet Courts tried a number of cases with extensive social implication and significant for rule making, which further defined the boundaries of cyberspace rights, code of conduct, and governance rules in accordance with the law.

(1) Clearly define the responsibility of network platforms. For example, the "WeChat Mini Program Infringement Case" tried by Hangzhou Internet Court clarified that WeChat Mini Program only provided architecture and basic access services and thus was not subject to the "notification-deletion" rule, which ensured the healthy development of new Internet business form.

(2) Effectively protect network intellectual property rights. For example, the "Artificial Intelligence Copyright Case" tried by Beijing Internet Court determined the way to protect the content generated by intelligent computer software.

(3) Strongly crack down on online infringements. For example, Hangzhou Internet Court declared that the commercial use of personal Credit Sesame data constituted a violation of privacy and clarified the legal responsibilities for abusing personal credit data.

(4) Regulate the development of emerging network industries in accordance with the law. Internet Courts have stepped up the crackdown on online gray and black markets, such as click farming and identity theft, and explored ownership determination and competition protection rules for emerging industries such as blockchain, Bitcoin, and big data. In addition, overseas parties highly approved of the "Peppa Pig Copyright Infringement Case" tried by Hangzhou Internet Court, which set an example for the international community and expanded China's international judicial influence on Internet cases.

7.5 Deepening Internet Law Promotion, with the Public's Cyber Legal Awareness Significantly Improved

The authority of the law comes from people's heartfelt support and true faith. To advance cyberspace governance by law, we must continuously promote laws and regulations and inspire all netizens to become faithful advocates, active followers and firm defenders of the socialist rule of law. In the past year, relevant departments have continuously strengthened and expanded the coverage of Internet law promotion and education, and effectively raised the public's cyber legal awareness, which has produced a broad consensus for realizing sustainable and healthy development of cybersecurity and informatization and building China's strength in cyberspace.

7.5.1 Actively Promoting Cyberspace Legal Education

7.5.1.1 Promoting Cyberspace Legal Education in China

In order to implement the legal education plan in the "Seventh Five-Year Plan" of the central government and fulfill the "Law Enforcer as Educator" requirement, the Cyberspace Administration of China carried out a series of cyberspace legal education across the country in recent years to improve legal and professional competence of cyberspace-system-related civil servants and corporate employees, and promote the concept and boost the awareness of the rule of law, which led to a nationwide study of cyberspace laws and regulations. From October to December 2018, the National Internet Information Office worked with the National Office of Law Popularization (under the Ministry of Justice) and organized the "National Cyberspace Law Promotion in Agencies and Companies" campaign. Through online and offline activities such as cyberspace legal education courses, visits to cyberspace firms, cyberspace knowledge contests, and cyberspace law promotional material competition, the campaign attracted more than 8,000 employees from cyberspace-system related agencies and enterprises, covering more than three million Internet users.

7.5.1.2   Organizing National Cybersecurity Promotion Week and Law Promotion Day

On September 19, 2019, China organized the Law Promotion Day during National Cybersecurity Promotion Week with various law promotion activities in different forms across the country. Through promotion panels and posters, promotion slogans, promotion and education materials, on-site consultation, official public accounts and other efforts, relevant agencies helped the public by promoting cybersecurity laws and regulations, exposing cyber crime tactics, and teaching precautions to prevent personal information leakage, and encouraged the public to actively report illegal online activities, all of which were well received.

7.5.2   Innovating Law Promotion and Education Activities

Local authorities actively carry out various forms of law promotion activities, guiding netizens to form right values and regulating online behaviors to shape a preferred environment for comprehensively promoting the rule of law in cyberspace and building a clean and healthy cyberspace.

7.5.2.1   Deepening Law Promotion at the Grassroots Level

Cyberspace law promotion shall focus on meeting the legal needs of people at the grassroots level and specifically keep deepening law promotion to the public at the grassroots level. In August 2019, Cyberspace Administration of Tianjin Municipal Party Committee invited experts, scholars and industrial representatives from universities, research institutes and big data companies as lecturers and organized a hundred lectures on "Learning Big Data" at government and party agencies at all levels, companies and institutions, schools, communities and townships, to solidly facilitate China's big data development strategy and promote and interpret *Regulations on Promoting the Development of Big Data in Tianjin*.

7.5.2.2   Focusing on Law Education for Minors

Minors are "Indigenous Netizens", so cyberspace law promotion shall start with minors, and focus on helping them get legal education, establish legal awareness and cultivate law-abiding habits. In 2019, the Cyberspace Administration of Hebei Provincial Party Committee and related departments organized the "Children's Day Special Program of Hebei Cyberspace Law Promotion–An Interesting Internet Law Promotion Class to Build Clean and Healthy Cyberspace for Better Growth". The program was broadcast live during the prime time of Hebei Traffic Radio and circulated among the public on new media terminals. Through cartoons, songs, cautionary cases, flash mobs and other forms that appealed to children, the program aimed to promote Internet laws and regulations and guide children to "surf and use the Internet in accordance with the law". The Cyberspace Administration of Hebei Provincial Party Committee also posted ads to recruit enthusiastic parents to build "Cyberspace Guardian Mothers", a cyberspace law promotion organization for minors in Hebei Province.

7.5.2.3 Organizing Knowledge Contests to Improve Cyber Competence

Cyberspace law promotion shall appeal to the public's mindset, and make full use of innovative approaches and methods such as knowledge contests to carry out targeted law promotion and improve its effectiveness. Xinjiang Uygur Autonomous Region Party Committee, Cyberspace Administration Office and other departments jointly organized the 2019 Autonomous Region Cyber Competence, Fire Safety and Traffic Safety Campus Promotion activities. Since April 2019, towns, prefectures and cities across the region held nearly 100 offline lectures and online knowledge contests with approximately 450,000 participants, with an average score of 92.56 points and a 90% passing rate. In addition, these activities also included safety classes that suited to the age of primary school students with vivid courseware, animation videos and other forms, which made it easier for students to understand and take part in the Q&A interaction.

In face of increasingly complex international landscape, it is necessary to accelerate the construction of the rule of law in cyberspace and safeguard national cyber sovereignty and interests. In face of risks and challenges brought about by the rapid development of new technologies and applications, it is necessary to speed up the construction of the rule of law in cyberspace, further clarify the rights and responsibilities of relevant subjects, regulate cyber behavior, and encourage healthy development. Only by tightening the web of law, strengthening legal power, and keeping coordinating social forces, balancing social interests, adjusting social relations and regulating social behavior in accordance with the law, can we ensure a truly vibrant and orderly cyberspace.

Chapter 8
International Governance, Exchange and Cooperation in Cyberspace

8.1 Outline

At present, a new round of technological and industrial revolution represented by information and communication technology has created new demands and posed new challenges for cyberspace governance. Unilateralism and protectionism are spreading, and international cyberspace governance has entered the deep water. There is a growing consensus among the international community to improve cyberspace governance mechanism, facilitate the formation of universally accepted international rules on cyberspace, and drive the reform of the global cyberspace governance system.

Since China's full-function access to the Internet in 1994, Internet has developed rapidly in China with world-renowned achievements, making it a significant Internet power. In recent years, China has actively participated in international cyberspace governance. President Xi Jinping's "Four Principles" for advancing the reform of the global Internet governance system and "Five Propositions" for building a community with a shared future in cyberspace have contributed Chinese solution and wisdom to the establishment of a more equitable and reasonable global Internet governance system. These are increasingly popular in international practice, forming a broad consensus among international community. Chinese government has thoroughly implemented the concept of international cyberspace governance proposed by President Jinping, actively carried out diversified and multi-level international cooperation in cyberspace, vigorously promoted the construction of the 21st Century Digital Silk Road, and strengthened bilateral, multilateral and regional dialogues and exchanges, which has further enhanced its voice and influence in international cyberspace.

© Publishing House of Electronics Industry 2021
Chinese Academy of Cyberspace Studies, *China Internet Development Report 2019*, https://doi.org/10.1007/978-981-33-6930-6_8

8.2   Background for China's International Cyberspace Governance

8.2.1   Growing Uncertainties in International Cyberspace Environment

Currently, the international environment for cyberspace is undergoing profound changes, and international cyberspace governance has entered an important transition stage. New technologies and applications such as artificial intelligence, 5G, Internet of Things, blockchain and quantum computing continue to emerge, exerting an important impact on the governance pattern. There still exist differences in the leading principles and models of international governance, making it difficult to form a joint effort to promote international cyberspace rule-making progress. Emerging economies and developing countries are accelerating the development of their digital economy, rapidly advancing their digital capabilities, and gradually increasing their international voice and influence. The government agencies of traditional Internet powers have further highlighted their role in cyberspace governance, which has brought new changes to the established order of the digital world. Geopolitical conflicts have spread to cyberspace, and the competition of comprehensive national strength is intensifying among major powers. The basic principles of international law such as sovereign equality, peaceful settlement, and non-interference with internal affairs of *Charter of the United Nations* are not effectively implemented in cyberspace. It's hard for the existing international governance mechanisms to adapt to the rapid changes of Internet development and international governance, which further adds to the fragility and uncertainty of cyberspace. Topics such as how to facilitate the reform of global Internet governance systems, how to ensure peaceful use of cyberspace and the harmonious interaction of international relations, how to consolidate the base for the order of mutual trust, fairness, and sharing in cyberspace, and how to realize the sustainable development of cyberspace, have become important issues for China and for countries around the world alike.

8.2.2   Opportunities and Challenges Coexisting in China's International Cyberspace Governance

In recent years, China seized the historical opportunities brought by the new round of technological and industrial revolution, focused on fostering an innovative application environment for Internet, made vigorous effort to build, develop and regulate Internet, and yielded world-renowned achievements. China actively participates in international cyberspace governance, pragmatically promotes international cooperation, and is committed to mutual benefit and common development of all parties. It has grown from a participant and beneficiary of Internet development to a constructor and contributor to international cyberspace governance, and continuously push it towards a more equitable and reasonable direction.

In the meantime, China faces many challenges with international cyberspace governance: the landscape of international governance is increasingly complex and the struggle among major powers in cyberspace is intense. Some countries even use information technology products and services as important means to fight and contain other countries, which add to the confrontation and fragmentation of cyberspace. Under such circumstances, China faces a significant increase in instability and uncertainty as it engages in global cyberspace governance. The rapid iteration of new technologies and applications continuously generates new governance issues and needs, posing new challenges to China's governance capabilities. Therefore, it is an important mission that China participates in international cyberspace governance, plays a greater role as a responsible cyber power, and enhances its voice and influence, to build a digital world of mutual trust and common governance with other countries.

8.3 China Actively Engaging in and Promoting International Cyberspace Governance

International community is highly concerned about cyberspace governance, and the limitations of existing international governance mechanisms are growing more prominent. The reform of the international cyberspace governance system is urgently in need of consensus. China conformed to the trend of the times and proposed that international cyberspace governance should insist on multilateral engagement with various parties and give full play to the role of various entities such as governments, international organizations, Internet companies, tech communities, non-governmental organizations, and individual citizens. In addition to promoting cyberspace governance under the UN framework, the active role of various non-state actors should be considered more. In November 2018, President Xi Jinping emphasized in his congratulatory letter to the Fifth World Internet Conference that although countries around the world varied in their national conditions, Internet development stages and practical challenges, they shared the same desire to promote the development of their digital economy, the same interests in response to cybersecurity challenges, and the same needs to strengthen cyberspace governance. All countries should deepen pragmatic cooperation, take common progress as the driving force and win-win as the goal, blaze a path of mutual trust and common governance, and invigorate a cyberspace community with a shared future. Over the past year, China actively facilitated the reform of global Internet governance system, and worked with the international community to promote the peaceful, secure, open and orderly development of global cyberspace.

8.3.1 Deeply Engaging in Important Platform Activities for International Cyberspace Governance

China has always advocated that the United Nations should play a major role in the formulation of international cyberspace rules, conduct consultations and dialogue on cyberspace rules with local international organizations, encouraged tech companies, technical communities, social organizations and research institutions to contribute to the establishment of standards and norms for technological innovation, and worked with various countries to build a multilateral, democratic and transparent global cyberspace governance system.

8.3.1.1   United Nations Internet Governance Forum (IGF)

China engaged deeply in the 13th United Nations Internet Governance Forum (IGF), and held a number of sessions on cyberspace trust and security, data privacy, artificial intelligence and other topics with 3,000 representatives from political, business, academia communities and non-governmental organizations around the world. The Cyberspace Administration of China, China Association for Science and Technology and other organizations held a number of events such as "Technological Innovation and the Evolution of Global Internet Governance Rules" at the forum, where they conducted in-depth exchanges on topics such as China's Internet policy and the evolution of global Internet governance rules, strengthened dialogue and cooperation with relevant organizations through IGF, fully delivered China's voice and philosophy, and achieved positive results.

8.3.1.2   World Summit on the Information Society (WSIS)

In April 2019, World Summit on the Information Society (WSIS) was held in Geneva, Switzerland. This was the largest annual summit in the field of information and communication technology organized by the United Nations. Chinese information and communication enterprises and research institutions actively participated in the relevant activities at the WSIS Forum, and all the 26 recommended projects were nominated for the WSIS Project Awards. Among them, China Unicom's "Data Encryption Leak-proof and Tamper-resistant Network System Based on Quantum Communication Trunk Line" won highest WSIS award, and projects recommended by China Mobile, China Communications Technology Co., Ltd., and Huawei were also honored. Relevant companies and research institutions held seminars on topics such as 5G, artificial intelligence, digital economy, cybersecurity, information accessibility and other fields at the summit. They introduced China's successful practices and solutions in related fields, and attracted great attention from all participants.

8.3.1.3 International Telecommunication Union (ITU)

China has worked closely with International Telecommunication Union (ITU) to promote project cooperation in the field of information and communications and increase international influence based on multilateral platforms. In April 2019, the Export-Import Bank of China and ITU signed a memorandum of understanding on strengthening cooperation in the digital field under the "Belt and Road" Initiative to promote the 2030 Agenda for Sustainable Development. The memorandum aimed to strengthen cooperation between the two parties, promote project implementation, and expand the application of information and communication technology in developing countries. In July 2019, at the 32nd meeting of the ITU Study Group 5 International Mobile Telecommunications Working Group, the Chinese delegation submitted a 5G NR wireless technology and actively promoted the introduction of the ITU 5G technological plan.

8.3.1.4 Internet Corporation for Assigned Names and Numbers (ICANN)

China actively participates in the work of Internet Corporation for Assigned Names and Numbers (ICANN), including with EAI Project Group, Root Server System Advisory Committee (RSSAC), and Root Zone Chinese Script Generation Panel and other groups in different fields with varying contribution. These efforts play an important and positive part in advancing openness and transparency of ICANN and promoting the fair distribution of basic Internet resources.

8.3.1.5 Forum of Incident Response and Security Teams (FIRST)

Since its establishment in 1990, Forum of Incident Response and Security Teams (FIRST) has promoted global cybersecurity incident response capabilities through technological development, standard formulation, and training & education. In October 2018, National Computer Network Emergency Response Technical Team/Coordination Center of China and FIRST co-hosted the Asia-Pacific Regional Conference in Shanghai, urging all parties in the Asia-Pacific region to further enhance coordination, enhance trust, strengthen cooperation and share information.

8.3.2   Building China-Leading International Cyberspace Governance Platform

8.3.2.1   Continuing to Promote "Wuzhen Process"

The World Internet Conference (Wuzhen Summit), an annual global Internet event initiated and organized by China, has successfully built itself as an international platform bridging China with the world and a Chinese platform for the sharing and co-governance of Internet. It aims to facilitate closer connection, more frequent exchange and deeper cooperation between countries in cyberspace. After 5 years of cultivation and development, the World Internet Conference has built up its influence, expanded its brand effect and boosted its leadership in the global Internet field. Over the past six years, through the advocacy and promotion of the World Internet Conference, the international community has strengthened dialogue, negotiation and cooperation in cyberspace, step up global network infrastructure construction, deepened multi-level digital economic cooperation, and continuously improved cybersecurity capabilities. Online cultural exchange and sharing are more frequent, and global cyberspace governance is moving towards a more equitable and reasonable direction. In November 2018, the Fifth World Internet Conference, themed "Creating a Digital World for Mutual Trust and Collective Governance—Towards a Community with a Shared Future in Cyberspace", achieved fruitful results in terms of exchange of ideas, theoretical innovation, technological display, and economic and trade cooperation. The conference's advisory committee of high-level experts successfully held its re-election and released *Wuzhen Outlook 2018*, the result produced at its annual meeting that summarized and looked forward on global Internet development and governance in five aspects: Internet development and innovation, cyber culture, digital economy, and cybersecurity and Internet governance, which vigorously promoted "Wuzhen Process".

8.3.2.2   Connecting with the Belt and Road Construction

The Belt and Road Initiative is an important platform for China to carry out international cooperation in cyberspace. In recent years, the new model of "Internet+" has provided new momentum for the rapid advancement of the Belt and Road Construction and the economic growth of related countries. Based on the philosophy of "openness, innovation, tolerance and inclusiveness", China vigorously promotes cooperation in the construction of the Digital Silk Road. At the "Digital Silk Road" Sub-Forum during the Second Belt and Road Summit Forum for International Cooperation held in April 2019, the participants focused on advancing innovation-driven development as well as the construction of digital economy, artificial intelligence and smart cities, and discussed about deepening cooperation in these fields. Representatives from various countries unanimously stated that pragmatic actions should be taken within the Belt and Road framework to promote the integration and development of digital economy and real economy, accelerate the transformation of new and old development drives, and create new industries and new business forms. They would also jointly facilitate the construction of information infrastructure, improve network interconnection and interoperability, and promote technological cooperation, openness and sharing.

At present, China has established a bilateral e-commerce cooperation mechanism with 17 countries along the Belt and Road to jointly build a large cross-border e-commerce platform, signed a digital silk road construction cooperation agreement with 16 countries, jointly launched the Initiative on Belt and Road Digital Economy Cooperation, built more than 30 cross-border terrestrial optical cables and more than 10 international submarine optical cables with countries along the route, explored telemedicine cooperation in more than 50 countries, and cooperated with related companies in more than 40 countries to develop new applications such as mobile payment.

8.3.2.3 Developing Emerging Internet Technologies and Innovating International Exchange Platforms

China attaches great importance to the technological development and innovative applications in emerging Internet technologies and digital economy, and actively joins forces with all parties to build open, integrated, shared, and win-win international exchange platforms to promote collaborative innovation between the scientific and technological circles and the business community. In particular, in the cutting-edge technology field, such as artificial intelligence, industrial Internet and virtual reality (VR), China has successively held a number of international conferences and events, including the China International Software Expo, the World Robot Conference, the Smart Industry International Expo, the World Artificial Intelligence Conference, the World Industrial Internet Conference, the World Intelligent Network Automobile Conference and the World Conference on VR Industry. While showing its ability in innovation, China also demonstrates that it is willing to share its Internet innovation with the international community more openly.

8.3.3 Actively Promoting Regional Multilateral Cooperation

8.3.3.1 Advocating the G20 Takes Responsibility for Digital Governance

At present, the world lacks momentum for economic growth, calling for efforts to narrow the digital divide and meet tough challenges of peace deficit, development deficit and governance deficit. With the rapid development of digital economy, issues such as data governance have become a topic of great concern to the G20. In

June 2019, at the 14th G20 Leaders' Summit held in Japan, Chinese President Xi Jinping emphasized that the G20 must persist in reform and innovation, tap growth momentum, keep up with the times and improve global governance. At the special meeting on digital economy, President Xi Jinping pointed out that G20 needed to jointly improve data governance rules to ensure the safe and orderly use of data, to promote the integrated development of digital economy and real economy, to strengthen the construction of digital infrastructure, and to promote interconnection. And G20 needed to boost the inclusiveness of digital economy and bridge the digital divide. As a major power in digital economy, China would actively participate in international cooperation and keep an open market to achieve mutual benefit and win-win outcome. These propositions were widely recognized and received active response from the participants.

8.3.3.2   Promoting the Inclusive Development of Asia-Pacific Digital Economy

As the highest-level, widest and most influential economic cooperation mechanism in the Asia-Pacific region, the Asia-Pacific Economic Cooperation (APEC) has attached great importance to issues related to the development of digital economy in recent years. China is an active advocate and firm practitioner of Asia-Pacific cooperation. As a major power in the development of digital economy, China has contributed a large number of innovative applications and achievements to the rapid development of the Asia-Pacific digital economy. In 2014, the APEC Conference in Beijing issued *The APEC Accord on Promoting Innovative Development, Economic Reform and Growth*, and brought Internet economy into APEC cooperation framework for the first time through *The APEC Initiative of Cooperation to Promote Internet Economy*. In November 2018, the APEC Conference in Port Moresby discussed digital economy and inclusive development under the theme of "Harnessing Inclusive Opportunities, Embracing the Digital Future". At the informal leaders' meeting, President Xi Jinping proposed to deepen cooperation in digital economy with all parties in the Asia-Pacific region, cultivate more common interests and economic growth points, and help members at different development stages share the fruits of digital economy development. At present, 35% of the economies in the Asia-Pacific region still face poor digital environment and underdeveloped technological capabilities, with difficulties such as lack of digital infrastructure, low Internet coverage, expensive network fees and insufficient supply of talents. Chinese companies are deeply involved in the digital infrastructure construction of countries in the Asia-Pacific region, providing e-commerce and financial technology services to local users, and helping local small-and-medium-sized enterprises onto the express train of digital economy development.

8.3.3.3   Deepening Digital Cooperation Among BRICS Countries

As an important multilateral cooperation mechanism represented by emerging markets and developing countries, the BRICS countries have agreed on a wide range of issues including strengthening future network research, promoting digital transformation and deepening cooperation in information and communication practices by improving systems, building consensus and coordinating overall planning. The BRICS is playing an increasingly important role in promoting the reform of the global Internet governance systems. China attaches great importance to cyber cooperation under the BRICS framework, actively strengthens cooperation in areas such as technological innovation, and deepens partnership with other BRICS members. In June 2019, President Xi Jinping emphasized at the BRICS Leaders' Meeting that the BRICS countries should deeply participate in global innovation cooperation, jointly advocate mutual benefit and win-win outcome, and create an open, fair, and non-discriminatory environment to encourage emerging markets and developing countries and their companies to take part in technological innovation and share benefits thereof. BRICS should also strengthen cooperation in digital economy and other fields to better withstand external risks. The meeting issued *The Press Communique of the BRICS Leaders' Osaka Meeting*, emphasizing the importance of interaction between trade and digital economy, calling on developing countries to participate more deeply in the global value chain, and promoting the digital transformation of BRICS countries. In August 2019, the Fifth BRICS Communications Ministers' Meeting was held in Brazil. China proposed initiatives to strengthen communication and connection, deepen open cooperation, and optimize market environment centering on the topic of promoting the digital transformation of the BRICS countries. In addition, the establishment of the Chinese branch of the BRICS Institute of Future Networks in Shenzhen will further strengthen cooperation of the BRICS countries in the field of information and communication.

8.3.3.4   Promoting Shanghai Cooperation Organization to Strengthen Cyberspace Cooperation

China attaches great importance to and fully participates in regional cyberspace governance under the framework of Shanghai Cooperation Organization (SCO), and actively promotes peace, stability, development and prosperity of regional and global cyberspace. In June 2019, at the 19th Meeting of the Council of Heads of State of Shanghai Cooperation Organization, President Xi Jinping delivered an important speech, proposing to insist on innovation-driven development in cyberspace and foster cooperative growth in digital economy, e-commerce, artificial intelligence, big data and other fields, which contributed Chinese wisdom to the further opening and cooperation of SCO in cyberspace. The participating heads of state jointly signed and issued *The Bishkek Declaration of the Council of Heads of State of the Shanghai Cooperation Organization*, and reached consensus on enhancing mutual trust and active cooperation and improving governance capabilities in the field of cyberspace.

China and other member states stated that they would crack down on the use of ICT to undermine political, economic and social security of SCO member states, combat the spread of terrorism, separatism and extremism through Internet, and oppose any discriminatory practices against the development of digital economy and ICT technology under all circumstances. The member states believe that it is necessary to formulate rules, principles, and norms of responsible national conduct in information space acceptable to all parties, actively cooperate to ensure information security in the SCO region, and call on all UN member states to further promote the development of responsible national conduct in information space.

8.4   Actively Carrying out Cyberspace Exchange and Cooperation

8.4.1   Properly Handling Sino-US Cyberspace Relations

The Sino-US relationship is one of the most important bilateral relations in real space and cyberspace, and has a pivotal impact on international cyberspace governance. Over the past few years, the United States has unilaterally emphasized "America First", pursued unilateralism and economic hegemonism, and discarded the basic international engagement principles of mutual respect and equal consultation. It attacked all sides and provoked international trade frictions, abused the concept of "national security" for trade protection, and continued to expand the scope of "long-arm jurisdiction", which covered many areas including cybersecurity. As the United States made major changes with its strategy on China, competition and friction between China and the United States in cyberspace intensified. The United States has imposed strict technological export controls on China, unilaterally suppressed Chinese technology companies such as ZTE and Huawei, and restricted the exchange of scientific and technological personnel between China and the United States. These measures have affected the normal exchanges between the industries, science and technology, academia, and the public in China and the United States, and further damaged the interests of all parties and seriously undermined the fairness and reasonableness of the international order in cyberspace. Faced with this situation, China still insists on negotiation based on equality, mutual benefit and integrity, advocates settling Sino-US issues through dialogue and consultation, and calls on both sides to carry out comprehensive and in-depth exchange in a frank manner. China proposes to properly handle differences and build Sino-US relations based on stability, cooperation and coordination together.

8.4.2 Expanding Sino-Russian Multi-level Cyber Communication and Cooperation

2019 marks the 70th anniversary of the establishment of diplomatic relations between China and Russia, and the relations between the two countries face new opportunities. In June 2019, the two countries signed *The Joint Statement on Elevating Bilateral Ties to a Comprehensive Strategic Partnership of Coordination for a New Era* and *The Joint Statement on Strengthening Global Strategic Stability between the People's Republic of China and Russian Federation*, stating both sides were committed to the development of a comprehensive strategic and cooperative partnership between China and Russia in the new era. The statement pointed out that both sides would expand exchanges in the field of cybersecurity, take further measures to maintain security and stability of their critical information infrastructure, strengthen exchanges in the field of cyberspace legislation, jointly promote the principles of Internet governance in compliance with international laws and domestic regulations, maintain peace and security of cyberspace based on equal participation of all countries, promote the construction of a global information and cyberspace governance order, and further facilitate the formulation of code of responsible national conduct in cyberspace under the UN framework and the formulation of universal legally binding documents to combat the abuse of information and communication technologies for criminal act.

Under the framework of strategic cooperation, the two countries have expanded multi-level cyber exchange and cooperation. In April 2019, the Third China-Russian Joint ICT Enterprise Workshop was held. Representatives from Chinese and Russian government departments, scientific research institutions, telecommunications operators, terminal manufacturers, Internet companies, software service companies and other departments discussed network infrastructure and Sino-Russian ICT corporate cooperation at the meeting. In June 2019, the China Internet Network Information Center (CNNIC) and the Russian ".RU" registration administration reached consensus on cybersecurity, internationalized domain name technology cooperation, application of emerging technologies, and mutual establishment of domain name resolvers as well as strengthening personnel exchanges in response to distributed denial of service (DDoS) attacks.

8.4.3 Keeping Deepening Sino-EU Cyberspace Cooperation

Cyberspace has become an important area of bilateral cooperation between China and Europe. In April 2019, the 21st China-EU Leaders' Meeting reached consensus on cyberspace governance and technological cooperation, emphasizing that China and the EU should maintain an open, secure, stable, accessible and peaceful ICT environment, continue to strengthen cyber exchanges and cooperation, and strive to promote the formulation and implementation of internationally accepted national codes of conduct in cyberspace within the framework of the United Nations. They should also strengthen cooperation in combating malicious activities in cyberspace through China-EU Cyber Task Force, including the cooperation in intellectual property protection, further consolidate dialogue and cooperation mechanism based on the 2015 China-EU 5G Joint Statement, and deepen practical cooperation in the field of 5G technology.

8.4.4 Strengthening Sino-French Cyberspace Governance Exchange and Cooperation

China and France have established broad consensus on major issues such as maintaining world peace, security and stability, protecting multilateralism and free trade, and supporting the active role of the United Nations. In March 2019, China and France issued a joint statement on protecting multilateralism and improving global governance, and reiterated that international law represented by *Charter of the United Nations* was applicable to cyberspace. Both sides are committed to promoting the establishment of universally accepted international norms on responsible behavior in cyberspace under the framework of the United Nations. The two countries will strengthen cooperation to combat cyber crime and terrorism and other malicious acts in cyberspace, and agree to continue to use the Sino-France dialogue on cyber affairs to strengthen relevant exchange and cooperation. China-France Global Governance Forum was held in Paris, France around the same time. The two countries exchanged views on promoting the innovative development of global digital economy and international cooperation in cyberspace and data security, and addressed the challenges from digital governance.

8.4.5 Strengthening Sino-UK Cooperation and Exchange on Internet and Digital Policies

China and the United Kingdom share a good foundation for cooperation in the Internet field. On April 9, 2019, the 7th China-UK Internet Roundtable co-sponsored by the Cyberspace Administration of China and the Department for Digital, Culture, Media and Sport (DCMS) of the United Kingdom was held in Beijing. The two sides discussed issues such as digital economy, cybersecurity, data and artificial intelligence, online child protection and corporate technological exchanges and cooperation, and reached consensus on cooperation in several fields. At the meeting, China and the United Kingdom jointly issued *Outcome Document of the Seventh China-UK Internet Roundtable*, agreed to strengthen cooperation and experience sharing in the field of Internet and digital policies, and reiterated that the "China-UK Internet Roundtable" would be held once a year.

8.4.6   Deepening Sino-German Internet Economic Exchange and Cooperation

China and Germany continue to deepen exchange and cooperation in the field of Internet economy. In June 2019, the Sino-German Dialogue on Internet Economy 2019 co-sponsored by the Cyberspace Administration of China and German Federal Ministry of Economic Affairs and Energy was held in Beijing. The two sides reached consensus on promoting win-win cooperation, opposing trade protectionism, maintaining peace and security of cyberspace, and vigorously developing digital economy. They jointly issued *Outcome Document of the Sino-German Dialogue on Internet Economy 2019*, agreed to strengthen regular exchange on economic legislation and regulatory framework for ICT technology at the government level, and emphasized their wish to continue to promote the development of bilateral economic relations. Both sides will strive to provide enterprises with a fair, equitable, and non-discriminatory business environment, and continue to cooperate on cybersecurity standardization.

8.4.7   Strengthening Sino-Italian Exchange and Cooperation on Digital Economy

Italy is the first Western developed country in the G7 to formally join the "Belt and Road" Initiative. The prospects of China and Italy jointly building the "Digital Silk Road" are promising. In October 2018, the China-Italy Digital Economy Dialogue was successfully held. Nearly 200 representatives from the governments, enterprises, think tanks and academic circles of the two countries participated in the seminar. The participants hoped that China and Italy would strengthen exchange and cooperation in the field of digital economy, continue to expand their cooperation on e-commerce big data, artificial intelligence, 5G network, smart city construction and other fields in the future.

8.4.8   Deepening Sino-Indian Internet Communication

China and India are both emerging economies. China's "Internet+ " Initiative and India's "Digital India" Project are highly complementary and hold great potential for cooperation. In September 2018, the 3rd China-India Internet Dialogue Conference was held in New Delhi, India. More than 600 representatives of Internet companies from China and India participated in the conference, one of the largest non-governmental business exchanges between China and India. Participants exchanged and shared their practices in more than 10 industry sectors such as venture capital, logistics, and e-commerce, and agreed that the cooperation between Chinese and

Indian Internet companies held broad prospects and Chinese investment would bring new development opportunities to the Indian Internet technology market. On the one hand, India possesses huge Internet market and holds huge potential in the field of Internet technology; on the other hand, the Indian Internet market is challenged by unbalanced development and small market size and in need of Chinese capital, technology and experience. All participants at the meeting are confident about the growing cooperation between Chinese and Indian on Internet companies.

25 years ago, China acquired full-function access to the Internet through a 64kb/s international cable, opening its Internet age. In the past 25 years, China built its own Internet from scratch, expanded it in size, strengthened its power, and caught the world's attention with stunning achievements. In face of complex and volatile international situation, China insists on moving forward with the world, keeping up with the time, doing the right thing, and seeking win-win solutions, always as a contributor to the development of Internet, a guardian of international cyberspace order, and a promoter of global Internet governance system reform, showing Chinese wisdom and responsibility on the world stage. China insists on taking common progress as the driving force and win-win as the goal, and working together with the international community to build a more vigorous cyberspace with a shared future to better benefit people all over the world through Internet!

Postscript

After 25 years of magnificent development, China has grown into a global Internet power. It has explored and forged a path of Internet governance with Chinese characteristics, and contributed Chinese experience and solutions to world Internet development. In 2019, guided by General Secretary Xi Jinping Thought on Socialism with Chinese Characteristics for a New Era, and his important thought on building China's strength in cyberspace in particular, China grasped the historical opportunities of informatization and made new major achievements in Internet development and governance. We hope that the compilation of *China Internet Development Report 2019* (hereinafter referred to as The Report) will contribute to promoting and interpreting General Secretary Xi Jinping Thought on Socialism with Chinese Characteristics for a New Era and his important thought on building China's strength in cyberspace in particular, fully picturing China Internet development, and systematically summarizing China Internet development and governance experience, so as to provide scientifically sound prospects and support China Internet development. It is also our hope that by examining China s Internet development, we could introduce China's experience and wisdom on Internet governance to countries around the world.

During the compilation of The Report, we received guidance and support from the Office of the Central Cyberspace Affairs Commission (hereinafter referred to as The Office). Leaders of The Office offered their insights, and all the affiliated agencies and units provided strong support for the preparation of The Report, especially with the provision of relevant data and material content. The Report is led and organized by Chinese Academy of Cyberspace Studies (CACS), and co-edited by National Computer Network and Information Security Administrative Center, China Academy of Information and Communications Technology, Peking University, Beijing University of Posts and Telecommunications, Institute of Information Engineering under Chinese Academy of Sciences, and other agencies. Main Contributors are Yang Shuzhen, Fang Xinxin, Hou Yunhao, Li Yuxiao, Li Changxi, Liu Shaowen, Feng Mingliang, Chao Baodong, Li Zhigao, Tian Yougui, Long Ningli, Tang Lei, Li Min, Liu Yan, Jiang Wei, Nan Ting, Zhao Yanwei, Han Yunjie, Dong Zhongbo, Wang Hailong, Li Bowen, Shen Yu, Li Xiaojiao, Wang Meng, Wang Xiaoshuai,

© Publishing House of Electronics Industry 2021

Chinese Academy of Cyberspace Studies, *China Internet Development Report 2019*, https://doi.org/10.1007/978-981-33-6930-6

Ma Teng, Zhao Gaohua, Xie Yi, Li Wei, Xu Xiu'an, He Bo, Jia Shuowei, Yang Xiaohan, Sun Luman, Tian Yuan, Yang Shuhang, Xiao Zheng, Song Shouyou, Wu Wei, Zhang Qiyuan, Gao Ke, Chen Jing, Yuan Xin, Xu Yanfei, Xu Yu, Li Yangchun, Deng Jueshuang, Cai Yang, Wang Zhongru, Yang Xuecheng, Sui Yue, Xie Xinzhou, Ding Li, Wang Xiaoqun, Xu Yuan, Meng Nan, Zhou Yang, Chen Kai, Mou Chunbo, Zhao Li, Jin Zhong, Wu Yanjun, Chong Dandan, Liu Shaohua, Li Shuai, Xin Yongfei, He Wei, Sun Ke, Zheng Anqi, Wang Mingzhu, Xu Ji, Hu Shiyang, Jin Jiangjun, Wang Lida, Lang Ping, Xie Yongjiang, Wang Wei, Liu Yue, Guo Feng, Fang Yu, etc.

The Report, though successfully published thanks to the strong support and considerable help from all sectors of society, is inadequate in terms of perspective and insight due to limited research experience and tight deadline. Therefore, we welcome valuable opinions and advice from government departments, international organizations, research institutes, Internet companies, and social organizations across different sectors, home and abroad, to help us produce better reports in the future and contribute more wisdom and strength to world Internet development.

Chinese Academy of Cyberspace Studies (CACS)

September 2019.

Printed in the United States
by Baker & Taylor Publisher Services